LONDON'S
INDUSTRIAL
HERITAGE

London's Industrial Heritage

Geoff Marshall

The
History
Press

First published 2013

The History Press
The Mill, Brimscombe Port
Stroud, Gloucestershire, GL5 2QG
www.thehistorypress.co.uk

Reprinted 2016

British Library Cataloguing in Publication Data.
A catalogue record for this book is available from the British Library.

ISBN 978 0 7524 8728 1

Typesetting and origination by The History Press

Contents

Preface

London is rarely considered an industrial centre. It has no coal mines, no great steelworks. And yet at the height of the Industrial Revolution between 1775 and 1800, apart from Lancashire with its vast cotton industry, the greatest concentration of Boulton & Watt steam engines was found in London. These engines kick-started the Industrial Revolution, and the presence of so many in London ably demonstrates the importance of the capital as an industrial centre. There were breweries by the score; the world's first plastic material was synthesised in London's East End, as was the first ever synthetic dye. After an extrovert Moravian demonstrated in Pall Mall that London's streets could be lit by gas, gasworks were built throughout the capital. There were power stations, sulphuric acid works, the River Thames was full of colliers bringing coal from Newcastle, and London's Docks were the envy of the world. Shipbuilding thrived on both sides of the Thames. Brunel's *Great Eastern* was launched from the Isle of Dogs, and at the Thames Iron Works at Bow Creek the world's first iron-hulled battleship, *Warrior*, was built. London was home to great engineers – Joseph Bramah invented his water closet and hydraulic pump, Henry Maudslay made machines to make machines, and John Rennie built his bridges. Motor cars, and even aeroplanes, were made. There were glassworks in Bankside, leather works in Bermondsey, and Royal Doulton pottery was crafted in Lambeth.

Londoners had to be fed and food was imported to the capital's markets – Smithfield, Billingsgate, Covent Garden, Spitalfields and Borough. There were (and are) food and drink manufacturers, and many household names began in London, such as Sainsbury's, Schweppes and Crosse & Blackwell. Huguenot refugees fled to London after the revocation of the Edict of Nantes and dominated the silk industry of Spitalfields. The match girls at Bryant & May went down with phossy jaw, and East End immigrants worked in the sweated labour of the textile industry. The list is endless.

For centuries men engaged in a common craft would combine together for their mutual benefit. Associations of this type were known as guilds and they dominated London's trade and industry throughout the medieval period and beyond. By the end of the fourteenth century, guilds were empowered to control wages and prices and had the right to inspect all goods and confiscate them if found faulty. However, the Industrial Revolution of the late eighteenth century and the increasing demand for goods severely reduced the influence of guilds. Their power was based on the organisation of a close-knit community. Once industry became more diversely spread around the country, the power of the guilds waned. Many survive, but their role has changed. Today they sponsor charitable causes and educational bodies and get involved in good work, often associated with their former trade. It is with the period following the demise of city guilds that the content of this book is mainly concerned.

As the book progressed I came to realise that more and more material could be included. The only sensible thing was to draw a line. Having said that, I appreciate that there are omissions and equally material has been included that should perhaps not be there.

Thanks are due to the many people who have helped in my research – the staff at the British Library, and the local studies libraries at Tower Hamlets, Newham and Southwark. I would like to give special thanks to Southwark Local History Library, for it was there that my interest in the subject was born when I was given the chance to contribute to their local history of Bankside. Thanks are also due to John Greenwood without whose *Industrial Archaeology and Industrial History of London: A Bibliography* this book would not have been possible.

Part 1
PUBLIC UTILITIES

1 The Electricity Industry

Electricity Generation

Michael Faraday's parents came from Yorkshire. They moved to London, where his father, James, worked as a blacksmith in Newington Butts. It was here – the area now known as Elephant and Castle – that Michael Faraday was born in 1791. In 1796, the family moved to rooms over a coach house in Jacob's Well Mews, Manchester Square. Faraday's education was rudimentary and in his own words, 'my hours out of school were passed at home and in the streets'. He was apprenticed to a bookbinder and his seven-year apprenticeship exposed him to books which he read avidly, most importantly those on science: 'I loved to read the scientific books which were under my hands … I made such experiments in chemistry as could be defrayed in their expense by a few pence per week.' In 1812, Faraday saved up sufficient money to attend the last four lectures that Sir Humphry Davy was giving at the Royal Institution. Faraday was entranced. He returned to the bookbinders, wrote up all the lectures and illustrated them with diagrams and sketches and bound them in a book which he sent to Davy, in effect asking for a job:

> My desire to escape from trade, which I thought vicious and selfish, and to enter into the service of science, which I had imagined made its pursuers amiable and liberal, induced me at last to take the bold and simple step of writing to Sir H. Davy expressing my wishes and a hope that, if an opportunity came in his way, he would favour my views: at the same time I sent the notes I had taken of his lectures.

After a while Davy took him on to clean and dust his apparatus at 25s a week and he never looked back. At the Royal Institution, on 17 October 1831, Faraday generated a 'wave of electricity' by moving a bar magnet into a coil of wire. Thus he discovered electromagnetic induction, the means to generate electricity by converting kinetic energy to electrical energy.

Due to the inadequacies of early dynamos, forty years elapsed before Faraday's discovery could be exploited. It was the semi-literate but brilliant electrical engineer Zénobe Théophile Gramme who opened the way for electricity to be generated on a commercial basis. Gramme was born in Belgium and in 1871 he demonstrated his dynamo, with ring-mounted armature, at the Academy of Sciences in Paris. The Gramme machine gave a much smoother supply of direct current (d.c.) than had been possible hitherto.

In its early days electricity was employed solely for lighting. The earliest form of lighting was the arc lamp. Humphry Davy, at the Royal Institution, had discovered the arc lamp in 1808. He allowed current (generated by electrolysis) to jump between carbon electrodes thereby producing a brilliant light. But early arc lamps had a short lifespan and were unreliable. It was not until the Russian telegraph engineer Paul Jablochkoff, working in Paris in 1876, invented the Jablochkoff Candle that arc lamps became a practical possibility. Jablochkoff's lamp consisted of two carbons rods separated by a paste of kaolin. When a current flowed, carbon paste between the two electrodes burned away emitting a dazzling light.

Given that both Gramme and Jablochkoff were working in France, it is hardly surprising that the French took the lead in installing electrically powered public lighting. In July 1878, the journal *The Electrician* highlighted the lack of progress in England, complaining, 'in London there is not one such light to be seen'. Yet within one month things were to change.

The first public building to be lit by the new electric light was the Gaiety Theatre, where arc lamps were installed in August 1878, described as 'half a dozen harvest moons shining at once in the Strand'. Then, a couple of months later, the French company Société Générale d'Électricité began an ambitious programme of street lighting along the Victoria Embankment between Westminster and Waterloo Bridges. Their power station – if that is the most appropriate way to describe it – was situated on the opposite side of the river, west of Charing Cross Bridge. It consisted of a wooden shed containing a steam engine, operating at 60psi pressure and driving a Gramme dynamo at 650rpm. Power lines ran beneath the Thames via a subway to power twenty Jablochkoff lamps on the north bank. More lamps were installed later, including some on Waterloo Bridge, but then, in 1884, the Jablochkoff Company went into liquidation and gas lights were reinstated.

Meanwhile, another scheme was under way at Holborn Viaduct. Sixteen Jablochkoff lamps were installed over a 500yd stretch of the viaduct and powered in a similar way to the Victoria Embankment scheme. *The Times* newspaper of 16 December 1878 reported:

> On Saturday evening the electric light was experimentally tried upon the Holborn Viaduct, at the instance of the City Commission of Sewers [and] the light was remarkably steady and brilliant. The trial will be continued for some time and arrangements are being made for it to light the Royal Exchange and the Mansion House.

The trial was, however, less than successful and Colonel Haywood, engineer to the City Commission of Sewers, was reported in *The Times* of 21 March 1879 as 'estimating the cost to be seven times that of gas and that the commission have resolved not to continue the experiment'. And that would appear to be that!

The gas companies were jubilant, boasting that 'we are quite satisfied that the electric light can never be applied indoors without the production of an offensive smell which undoubtedly causes headaches and in its naked state it can never be used in a room of even a large size without damage to sight'. The problem lay with the intensity and dazzling light of arc lamps, and the fact that they smelled and emitted a hissing noise. But the gas companies had not bargained with the pioneering work of Thomas Edison and Joseph Swan.

Joseph Swan was born in Sunderland in 1828. Having trained and set up in business as a pharmacist, he began experimenting with incandescent lamps in the mid-nineteenth century but was hampered because he could not obtain a good vacuum in his bulbs. Vacuum pump technology improved later in the century and, in 1875, he was successful in making an incandescent lamp with a carbonised thread as filament. He patented his bulb in 1878, just before Thomas Edison in America did just the same. The two men joined forces and the Edison & Swan Electric Light Company was founded. By not having to rely on arc lamps, electric light now became a practical possibility.

Thomas Edison was quick to exploit the new electric light. In January 1882, he got in touch with the City Corporation and proposed that the Holborn Viaduct scheme be revisited. Edison offered to light the viaduct free of charge for three months and also to supply private consumers. *The Times* of 13 April 1882 reported:

> From Newgate Street westward, across Holborn Viaduct, to Hatton Garden, the street and most of the buildings on either side of the street are now and for the next three months, will continue to be lit by Edison incandescent lamps. For the purpose of street lighting two of the incandescent lamps of 32 candle power each have been placed in every lamp-post and it had been hoped that last night permission would have been obtained from Colonel Heywood [*sic*] for the gas to be turned out in order that the effectiveness … might be proved the more satisfactorily … it is hoped that tonight the public will be able to judge on the matter themselves. Those that can obtain permission to see the machinery and appliances by which the electricity is generated and distributed will find a satisfactory answer.

The machinery *The Times* spoke of was sited at 57 Holborn Viaduct (on the north side of the road) and can claim to be the first power station in the world. Current at 100V d.c. was supplied from an Edison dynamo driven by a steam engine, with steam raised from a water tube Babcock & Wilcox boiler. One thousand Edison lamps of 16cp each were installed; later, a further 1,200 were powered from another generator at 35 Snow Hill. Along the route was the City Temple, which can lay claim to being the first church to be lit by electricity. The mains were copper conductors, encased in insulation within wrought-iron pipes. Holborn Viaduct Power Station continued to operate until 1886.

A Myriad of Small Stations

In the early 1880s, the Great Western Railway built a power station to light Paddington Station, its offices and the Great Western Hotel. According to *The Electrician*, it supplied 'by far the largest installation of mixed lighting hitherto made'. The power station stood a quarter of a mile from Paddington Station on the south side of the track. There were three 350kW alternators housed in a wooden building with double walls, so constructed to keep the noise down. It plainly did not, because in 1885 the residents of nearby Gloucester Terrace complained to the magistrates at Marylebone Police Court that the 'tremendous vibration and noise, added to the fumes of smoke and steam and dirt caused by the machinery, produced such a nuisance as to be almost unbearable'.

Meanwhile, in 1887, the Cadogan Electric Lighting Co. installed a series of storage batteries in the houses of residents in Chelsea, South Kensington and Knightsbridge, and, via overhead transmission lines strung from poles, supplied direct current from a small power station in Manor Street (now Chesil Court) next to Albert Bridge. The scheme did not prosper; by 1895, only twenty-five houses were being supplied. The Cadogan Co. was then taken over by the Chelsea Electricity Supply Co. which had obtained consent to lay mains beneath the street. Their power station, in the basement of a house in Draycott Place/Cadogan Gardens, supplied direct current to batteries in a series of substations, one at Draycott Place, the others at Pavilion Road and Egerton Gardens Mews. The substations were charged in series and supplied low tension d.c. to customers in parallel. Another power station was built in 1894 at Flood Street, and by 1911 the company's capacity had increased to 3,400kW, with further substations at Elm Park Gardens, Clabon Mews and Pond Place. The Chelsea Co. ceased generating in 1928.

The Whitehall Electric Supply Co. was formed in 1887 with the intention of lighting Whitehall Court (now the Royal Horseguards Hotel). A power station was built in front, underneath the road, and its customers included the church of St Martin in the Fields and various premises in Northumberland Avenue. Within the year it was taken over for £40,000 by the Metropolitan Electric Supply Co., which soon purchased a small power station in Rathbone Place, off Oxford Street. The Whitehall Court scheme supplied direct current, but at Rathbone Place the company opted for alternating current (a.c.). Another power station was constructed in Sardinia Street at the south-west corner of Lincoln's Inn Fields, and the Metropolitan Electric Supply Co. now served Marylebone, Bloomsbury, Lincoln's Inn and Covent Garden. A further station was built at Manchester Square in 1890. However, because of complaints – houses vibrated and clocks stopped – an injunction was served on the company. It was on the point of shutting down but the day was saved by replacing the noisy reciprocating Willans engines with three 350kW Parsons turbo-alternators, the first to be installed in the capital.

An early generating station was built, just south of Kensington High Street, by R.E. Crompton & Co. to illuminate Kensington Court. By 1890, the Kensington Court Electric Light Co. was supplying d.c. to the surrounding area. It was soon acquired by the Kensington & Knightsbridge Electric Lighting Co., which had a power station in Cheval Place. In 1892, 645kW was generated at Kensington Court and 410kW at Cheval Place.

The Westminster Electric Supply Corporation had two plants in 1890: one at Stoneyard, Millbank, near the House of Lords, supplying the Palace of Westminster; and the other at Chapel Mews, near St James's Park Station. They had further stations, all d.c., in Dacre Street, in Eccleston Place to supply Belgravia, and Davies Street to supply Mayfair. In 1904, a substation was built at Duke Street, Mayfair, and in 1910 the Millbank plant was demolished to make way for Victoria Tower Gardens. Meanwhile, a new station was built in Horseferry Road, near Lambeth Bridge.

The St James's and Pall Mall Electric Lighting Co. built their first power station in Mason's Yard, Duke Street. Sited in the heart of St James's, the station was prone to pollute the surrounding area with an oily spray which did nothing to please the members of London's fashionable gentlemen's clubs. As one of the engineers commented, 'it was almost a daily occurrence to see a gesticulating man pointing out his damaged top hat'. Wyndham's Club even obtained an injunction against the plant but then had the nerve to insist that their supply shouldn't be cut off! The company laid its mains in cast-iron culverts and on one occasion the outer casing became live. An old and unsuspecting horse was unfortunate enough to place his iron-clad hoof on the casing and received a fatal shock. The owner was compensated with £40 but soon afterwards a similar accident happened to a horse drawing a hansom cab. From then on it was commonplace for cab drivers to pass along Jermyn Street with old and worn-out nags in the hope that a similar fate might befall them. In 1893, a new station was constructed in Carnaby Street and by the end of the century the company amalgamated with the Westminster Company and began to build a new plant at Grove Road.

The Charing Cross Electricity Supply Co. traced its roots back to a small power plant installed in 1883 in the basement of the Adelaide Restaurant in the Strand owned by the Gatti Brothers. Two years later Messrs Gatti were supplying the Adelphi Theatre and in 1888 built a new station in Bull Inn Court, between Maiden Lane and the Strand. Acquired by the Charing Cross Co. in 1889, the new company soon expanded south of the river and built a plant between Waterloo and Blackfriars Bridges. In the early twentieth century consent was obtained to supply the City from a new power station at Bow. By 1919, its capacity was 74MW.

Originally known as the House-to-House Electric Light Supply Co., the Brompton & Kensington Electricity Supply Co. had an a.c. power station in Richmond Road, Brompton, in 1889. To begin with, every customer had a transformer in their own home before the company installed a series of their own transformers to supply low voltage. In 1928, when they finally ceased generation, the plant had a capacity of 8MW.

The County of London Electric Lighting Co. ran two power stations: one at the City Road basin of the Regent's Canal; the other on the banks of the Thames at Wandsworth. They began operation in 1896.

The first local authority to supply electricity in London was St Pancras. A power station was built in Stanhope Street, just to the east of Regent's Park. A second station opened in King's Road and was unique in that it used the hot gases from a refuse destructor to heat its boilers – an early example of energy conservation. Hampstead opened a small station on Finchley Road in 1893, to be followed by Islington which built a generating plant at Eden Grove, off Holloway Road. Ealing was soon to follow the example of St Pancras and harness heat from a waste destructor plant, and Shoreditch followed suit with a plant in Coronet Street, opened by Lord Kelvin in 1897. It came in for much criticism and was labelled a waste of time, but the chairman of the Shoreditch company insisted in a letter to the technical press: 'We are absolutely raising from our ashbin refuse sufficient steam to drive our electrical plant, giving a maximum output at our heavy load of 250kW, and this we are raising solely from ashbin refuse.'

Towards an Integrated Supply

So why did the electricity supply industry develop in such a haphazard way in London, and everywhere else for that matter? The answer is found in the early legislation to which the industry was subject. A select committee was established to look into the matter, chaired by Sir Lyon Playfair in 1879, and set the seal on how the industry was to develop for at least the next forty years. The committee recommended that electricity undertakings should supply power only within the area under the jurisdiction of their particular municipal authority. The result was scores of small power stations, each supplying only a small area.

The recommendations were given the authority of law by Joseph Chamberlain's Electric Lighting Act, which received its royal assent in 1882. The Act sought to prevent monopolies and also favoured electricity supply being put in the hands of municipal authorities by empowering them to compulsorily purchase private undertakings after a period of, first, twenty-one years and then (by the terms of a later Act) forty-two years. There were thus conflicting interests between local authorities and private concerns. The Electric Lighting Act of 1909 went some way to improve matters by allowing the Board of Trade powers (over the heads of local authorities) to authorise the breaking-up of streets and the compulsory purchase of land for building power stations.

The early legislation of the electricity supply industry has come in for much subsequent criticism – often laid at the hands of Joseph Chamberlain – whose horror of monopolies, and tenure as mayor of the thriving industrial city of Birmingham, no doubt influenced his preference for supply being provided by municipal authorities. But it was not only legislation that caused confusion – there were technical matters as well. These centred on whether it was preferable to supply direct current or alternating current. The controversy became known as the 'battle of the systems', with eminent men in both camps. Ferranti was a fervent advocate of a.c., whereas R.E. Crompton and indeed Thomas Edison favoured d.c. Alternating current supply eventually won the day because of its ability to transmit high voltages from large power stations over large areas without power loss. But in the early days of power generation d.c. offered many advantages. The most important of these was that it ensured security of supply because batteries could be used as a stand-by supply in the eventuality of plant breakdown. In contrast, if an a.c. generator broke down, the lights went out at once – a common occurrence which did much to enhance the cause of d.c. supply. It is little surprise, therefore, that the great majority of early power stations generated direct current – only the London Electric Supply Co. and the Metropolitan Electric Supply Co. favoured alternating current. But gradually advances in engineering – particularly the replacement of belt-driven generators by the turbo-alternator – began to favour large-scale a.c. transmission.

But there was still a multitude of different companies: in 1921, there were eighty separate supply concerns in London, supplying power from seventy different power stations, with fifty systems of supply, twenty-four voltages and ten frequencies. Looking at it from the customers' point of view, it meant that moving home meant changing all one's electrical appliances! And from the suppliers' viewpoint, because the stations were not interconnected each company was forced to install excessive reserve plant to cover for breakdown, or outage due to essential maintenance. The net result was that electricity was expensive!

The government responded later in the 1920s by setting up a committee under Lord Weir to enquire into the industry and make recommendations. Reporting in 1925, Weir recommended a national 'gridiron' of high voltage transmission lines taking power from the most efficient 'selected' power stations. His recommendations became the basis for the Electricity Supply Act of 1926. Selected stations – still either in private hands or owned by municipal authorities – generated alternating current at a standard frequency and sold it to the newly formed Central Electricity Board (CEB). Each company then repurchased electricity from the grid at a price not greater than it would have cost them to produce had the grid not been in operation. They then supplied their own customers in the same way as before. The CEB had the job of selecting stations and forcing inefficient undertakings to close their power plant and purchase supply from the grid. For the consumer, the result was a halving in the price of electricity.

The Second World War curtailed the construction of new power plant and, combined with a shortage of coal, power cuts after the war were inevitable. The building of new and large efficient power stations commenced immediately after the war. In London a new Bankside station was begun.

The post-war Attlee government nationalised the industry and by the terms of the Electricity Act of 1947 the British Electricity Authority was formed, later to be replaced by the Central Electricity Board and then the Central Electricity Generating Board (CEGB) in 1958. The CEGB ran all power stations in England and Wales and sold power to twelve area boards; in London this was the London Electricity Board (LEB). In the early 1990s, the industry was privatised but by that time all London's power stations had shut down and supply was taken from more efficient stations elsewhere.

Famous Thames-side Power Stations

Deptford Power Station – Ferranti's Dream
The story of Deptford Power Station starts in the most unlikely of places – New Bond Street in London's fashionable West End. There, in the mid-1880s, Sir Coutts Lindsay and Lord Wantage put up the money to illuminate the Grosvenor Gallery by means of a steam engine driving two Siemens alternators generating current at 200V. So impressed were local shopkeepers and residents that they asked to be connected to the system. Accordingly, overhead lines were strung along iron poles from house to house and transformers installed in each property to reduce the voltage. Rather than installing meters, each consumer paid £1 per year for every 10cp lamp and £2 for every 20cp one.

As with all new ventures, problems arose and so the young and ambitious Sebastian Ziani de Ferranti was called in to solve them. Ferranti was descended from a noble Italian family and was educated at St Augustine's Roman Catholic College in Ramsgate. After the briefest apprenticeship at Siemens, he set up in business at the tender age of 18 with the engineer Alfred Thompson and the lawyer Francis Ince. Ferranti immediately set to work to update the Grosvenor Gallery station by replacing the Siemens alternators with ones of his own. But Ferranti and his backers had other, more ambitious, schemes in mind. A new company was set up, the London Electric Supply Corporation (LESCo), with an authorised capital of £1 million in shares of £5 each.

Ferranti's dream was to supply large areas of London from what was, by the standards of the day, a giant power station situated by the river. There would therefore be an ample supply of cooling water, and coal could be delivered at minimum cost by sea from the coalfields in South Wales or Tyneside. He took his lead from the example of the gas companies, such as the Gas Light and Coke Co. which had consolidated their works at the remote site of Beckton. To quote Ferranti:

> The business of distributing electrical energy must be done on a large scale to be commercial, and to attain this we must supply a large area ... and we must do this from a site not in the congested heart of a big city but from a position best suited by its natural advantages to the carrying on of such an undertaking.

The site chosen was Deptford and LESCo set about building their station on land just to the west of Deptford Creek on a 3-acre site called the Stowage, previously used by the East India Company. Initial plans were for four 10,000hp engines, powered by steam from eighty boilers, driving four alternators supplying current at the then unheard of voltage of 10,000V.

Ferranti was newly married, but the building of Deptford Power Station was still at the forefront of his mind – he would often stay all night at the plant. His wife was to write later: 'The first thing I remember during those first months of married life was Deptford, and again Deptford. We talked Deptford and dreamed Deptford.' Ferranti's obsession did not go unnoticed: the journal *The Electrical Engineer* called him the 'Michelangelo of that installation because from first to last, from foundation to highest turret … all were specified or designed by one man, and the credit of that success will have to be given … to Ferranti'.

The first transmission cables from Deptford to supply London proved inadequate and so Ferranti, true to his character, set about designing and manufacturing them for himself. Rather than dig up streets and lay the cables underground, he came to an arrangement with the local railway companies and laid his cables along the railway track running from Deptford into London via Cannon Street, Blackfriars and Charing Cross railway bridges. He came to a similar agreement with the Metropolitan & District Underground Railway.

But the Board of Trade concluded that four 10,000hp engines in one location at Deptford would be a risk to the continuity of supply should there be a breakdown. The Board was also keen to ensure competition and recommended both a.c. and d.c. distribution. The consequence was that Deptford's area of supply was halved. In response, LESCo modified their plans and installed twenty-four Babcock & Wilcox boilers, producing 414,000lb of steam per hour driving two 1,250hp Corliss steam engines connected from fly wheel to the alternators. Then, on 15 November 1890, disaster struck at Grosvenor Gallery. Due to operator error, a momentary 5,000V arc was allowed to start a serious fire which eventually shut down the entire station. Supply was curtailed for three months and, not surprisingly, many customers transferred their allegiance elsewhere. LESCo lost money and many were quick to point the finger at Ferranti's wild scheme. The company was forced to cancel its order for the 10,000hp machines and in August 1891, Ferranti left Deptford. However, as we have said, his vision of large power stations supplying large areas of population was to be realised – but much later.

Electricity continued to be supplied from Deptford. In 1904, two 2MW alternators were installed to power the London County Council's tramways and later the station supplied the newly electrified railway. In time, a further station was built, Deptford West, and, in 1932, Deptford claimed to be 'the most efficient generating plant in the country'. Deptford continued to supply power until its final closure in 1983.

Battersea Power Station – Londoners Love It

Ten small companies combined in 1925 to become the London Power Co. Their first project was to build a giant power station on a 15-acre site with an eventual capacity of 400MW. It was Battersea Power Station.

Final consent was granted on 27 November 1927, but not before a storm of protest had erupted. Objections came from all quarters. Cosmo Lang, Archbishop of Canterbury, railed that 'it would cast the blight of soot and sulphurous fumes

Battersea Power Station.

on London's parks and gardens, picture galleries and oldest and noblest build-ings. How can a sense of civic beauty survive the progress of civilisation which is making a desert of the past and a dust heap of the future?' Lang was not alone in his outrage. He was joined by the King's physician, Lord Dawson of Penn; by both president and past president of the Royal Institute of British Architects; and, predictably, by the newspapers, who joined in the furore with headlines warning of 'a Blight of Poisonous Fumes', 'Noxious Gas by the Ton', 'Parks and Art Treasures in Danger' and 'Perils of Public Health'. The main sticking point was the vast amount of sulphur dioxide that would be emitted from the sta-tion, estimated at more than 3 tons per hour. In consequence, the Electricity Commissioners insisted that measures be taken 'for preventing as far as reasonably practicable the evolution of oxides of sulphur and generally for preventing any nuisance'. The solution was the incorporation of gas washing. In June 1933, two 69MW Metropolitan-Vickers units were installed and two years later a 105MW set, which remained the largest unit in the country until 1956.

The original intention of having nine metal chimneys was abandoned. Battersea's four upright chimneys are now a well-loved landmark, but to begin with – before the station was enlarged – there were only two. In response to the protests, Sir Giles Gilbert Scott was engaged as architect. Coming from a family of eminent architects, Scott is best known for his design of Liverpool's Anglican Cathedral, the elegant and graceful Waterloo Bridge and the red telephone kiosk. Scott was so successful that his design was given second place in a survey of modern buildings carried out by *The Architects' Journal* in 1939. Seemingly, he disliked the inverted table design but matters had gone too far and he was, much to the benefit of later generations, stuck with it. Scott's design is a steel-framed building clad in brick. No expense was spared: the control room was decorated

with Italian marble and parquet flooring, and the turbine hall was so clean and
well polished that the station superintendent commented that 'you could literally
eat off it'. The gas washing plant was the first of its kind in the world. Scrubbers
were fitted and the effluent gases sprayed with a chalk suspension. But the gas
washing plant was not without its unfortunate side-effects. A white plume was
formed, clearly visible to enemy aircraft, which could then use it as a marker to
bomb London. The liquid effluent was also polluting the Thames. The upshot was
that the gas washing plant was taken out of service. After the Second World War,
a second station was put in, Battersea B, with 100MW and 60MW sets, giving a
total capacity of 509MW. The famous station was now complete with its four distinctive chimneys. The design at Battersea, in fact, became the norm for all later
power stations in Britain in the immediate post-war years.

Electricity generation continued at Battersea until 1983 and there were many
innovations. In 1952, pulverised fuel was used for the first time, and the original
A station provided district heating for 30 acres of housing in Pimlico on the other
side of the river. Spent steam was passed beneath the river through 12in pipes to
a glass accumulator tower and gave heat to 2,000 flats – a total of 11,000 people.

Despite a new lease of life after the oil crisis of 1973, Battersea was supplanted
by more efficient stations in the Midlands and the North, and eventually closed
in 1983.

Bankside

The generation of electricity from Bankside began in the early 1890s with the
formation of the City of London Electric Lighting Co., registered in July 1891
with a capital of £800,000. There were two pairs of 25kW Brush arc lighters
and two 100kW alternators. They were driven by belting from Brush engines,

City of London Electric Lighting Co. Power Station, Bankside, c. 1891.

Boiler house at City of London Electric Lighting Co. Power Station, Bankside, *c.* 1901.

Turbine hall, City of London Electric Lighting Co. Power Station, Bankside, in the 1930s.

powered by steam from Babcock & Wilcox boilers. By 1901, there were a total of thirty-eight such boilers and capacity had risen to 10,500kW. Bankside was now the largest establishment of its kind in the world.

Noise and vibration were major problems with power stations. The vibration at Bankside was such that the men operating within the engine room were said to require 'sea legs' to work there and the proud statue of St George and the Dragon that stood in front of the building soon disintegrated and fell down. But the riverside site was seen as a positive advantage. There was a plentiful supply of water for cooling, ash could be removed easily by boat, and coal could be delivered by barge after offloading from seagoing vessels downstream at Blackwall.

In the early days of operation it was difficult to cater for rapidly changing demands for power. London fogs were an ever-present headache. Within a few minutes of the onset of a London 'pea souper', demand for power could rise dramatically and then equally quickly fall again when the fog lifted. The company tried to get knowledge beforehand of the coming of fogs by arranging for notice of them to be phoned through from the docks to the east (given that, it was reasoned, fogs came from this direction). Messages tended to arrive after the fogs had lifted and this instigated a policy of having a lookout posted high up on an observation platform around one of the chimneys.

The turbine came to Bankside in 1910 with a Parsons 2,500kW turbo-alternator replacing some of the earlier engine-driven machines. More quickly followed and by 1915 capacity had risen to 34,500kW. The eventual plant capacity for pre-war Bankside was 85,000kW, with two 15MW, five 10MW and one 5MW sets operating at a steam pressure of 270psi.

The original Bankside Power Station is now long gone but a remarkable incident occurred during its heyday. A 12-year-old boy was playing by the river water suction pumps when he lost his footing and fell in. A workman saw the accident and immediately raised the alarm and the pump was switched off. The boy, meanwhile, had been thundered, totally submerged, for well over a minute through the 3ft diameter entry pipe, through a couple of hairpin bends and along a distance of 50yd to a screened chamber under vacuum. The vacuum was turned off but it was twenty minutes before workmen were able to lift the cover of the chamber and perform what they thought would be the gruesome task of removing the body of a dead child. To their astonishment, when the cover was removed a loud cry was heard and there before the workmen's eyes was the adventurous lad, clinging to the screen with his head just out of the water. After the briefest of hospital visits he was pronounced 'none the worse for his experience' and sent home wrapped in a blanket.

By early 1939 it was realised that a new power station was needed, but the war intervened and plans were put on hold. After the war the shortage of power reached crisis proportions and permission was sought to alleviate this in 1946 by building a conventional coal-fired station. There were immediate objections and a public enquiry was held for their consideration. Opposition concentrated on two issues. First, that the power station would conflict with and hamper the view of St Paul's Cathedral, its vista now newly opened up by the wartime bombing. It was also anticipated that the sulphurous fumes emitted by the stack would drift across the river and damage the cathedral's stonework. Secondly, it was thought that the station would not fit in with the London County Council's 'County of London Plan', published in 1943. Its South Bank scheme envisaged a park for Bankside, with commercial buildings behind. Prominent in opposition were the London County Council (LCC), the City Corporation, Southwark Borough Council and the Dean and Chapter of St Paul's Cathedral. It was all to no avail and permission was granted by Lewis Silkin, Minister for Town and Country Planning, with the proviso that oil be used to fire the station instead of coal and that the flue gases be washed to prevent the emission of sulphurous fumes.

Fashions change and it is now generally recognised that Bankside Power Station (now Tate Modern) is a fine building and an integral part of London's architectural heritage. But at the time of its proposed construction, opinions were to the contrary and indeed if we examine views at that time, we see we are fortunate that the power station and art gallery-to-be were built at all.

Argument was heated. The Dean of St Paul's wrote to *The Times* on 16 April 1947:

> We submit that in the replanning of central London there is nothing more deserving of attention than the picture of London from the river and that to minimise the central and grandest feature in this picture would be to spoil an immemorial aspect of London and would be an irreparable mistake for which future generations would rightly blame those responsible. We earnestly hope therefore that the Government will assent to the views of the LCC, the City Corporation and Southwark Borough Council which are united in their opposition to this proposal.

The issues were debated in both the House of Commons and the House of Lords. These views must be seen alongside the fact that in the previous winter the country had been beset with power cut after power cut, and this was what was behind all the government's thinking. The Lord Chancellor, Viscount Jowitt, speaking on behalf of the government, summed it up: 'It would be poor consolation to a man rubbing his cold hands together to keep warm, to think, having sat there cold during two winters: "Well, I may be cold, but thank heaven the historical dominance of St Paul's over the river has been preserved."' Lord Llewellin wanted 'vistas both to and from St Paul's so that the cathedral might stand forth and be seen as the Arc de Triomphe can be seen down the Champs-Élysées' and he compared the power station proposal 'to introducing an alligator into the water lily pond'. Early to appreciate the beauty of power stations, Viscount Jowitt was quick to respond: 'On the contrary, I would say that this may well be another large and beautiful lily in the pond.' Poor Lord Llewellin could only reply: 'I must say I have never seen a water lily so well gilded as it has been by the noble and learned Viscount this afternoon. In fact I came to the conclusion that if we took for granted all the Lord Chancellor said we ought to go out and clamour to have

Demolition of City of London Electric Lighting Co. Power Station and construction of Bankside Power Station, c. 1959.

a further power station in Parliament Square. Let us have one next to Westminster Abbey so that we may all have the benefit of this wonderful structure he has described to us.'

In the Commons, debate proceeded along similar lines. One member wanted the power station to be built underground: 'It is quite easy to construct an underground railway; why not a power station?' Other gems included: 'We have been told that a power station can, in its way, be as beautiful as a cathedral. A hyena can be as melodious as a nightingale, but we do not like the way and prefer the nightingale.' And then there was: 'A power station is very much like a purgative pill. Put as much sugar on the outside as you wish but it is still a pill inside and it is no good putting fat boys and bulbous ladies on the outside of the building, it will still be a power station,' as well as 'Does anyone believe that the people of Rome would tolerate the erection of this power station in the shadow of St Peter's? Of course not.'

And so it was at a press conference at the premises of the City of London Electric Lighting Co. in Aldersgate Street on 19 May 1947 that the station's architect, Sir Giles Gilbert Scott, unveiled a model of the power station. The design had been amended to incorporate only one chimney as opposed to the previous two. 'Well, there is the so-called monster,' proclaimed Scott. He went on: 'I prefer this to the mountainous groups of offices as were proposed in the County of London plan. This station, with its slender tower or campanile, completes a composition and it lends itself without artificiality as a centrepiece. Why power stations should be considered as untouchables I cannot say. Power stations can be fine buildings.'

Scott recognised that power stations can indeed be beautiful buildings and this argument was later taken up by the architectural historian Gavin Stamp, who wrote in praise of Bankside. The term 'brick cathedrals' was coined, and Stamp described Bankside as 'urbane and elegant'. He also made the point that 'rebuilding the City with characterless sixties boxes made a nonsense of all the worthy talk of preserving the visual dominance of St Paul's'.

It was inevitable that consent should come, and work commenced in 1948. In the initial stages two turbines of 60MW each were installed. Later, in 1959, the original power station of the City of London Electric Lighting Co. was closed and demolished. Work started the next day to make way for the second half of the station with the installation of a further 60MW generator and a much larger 120MW unit.

There were three special features that set Bankside Power Station apart. First, its appearance, as a true 'Temple of Power', was the culmination of Sir Giles Gilbert Scott's power station designs. His brief was to design a building that harmonised with those existing and planned for the South Bank redevelopment. Secondly, to avoid dust and particulate smoke pollution, Bankside was the first large generating station to be fired solely by oil. And, thirdly, to complete the anti-pollution measures, a flue gas washing plant was incorporated to remove sulphur dioxide and other acid gases. Oil was transferred to the station from 500-ton tankers at an island jetty on the river front. Here it was pumped to three enormous mild steel tanks, set beneath the ground and to the south of the station. Each was 92ft in length and 24ft deep with a capacity of 4,000 tons.

The fuel oil could contain up to 4 per cent sulphur, which when burnt converted to acidic sulphur dioxide. In order to protect surrounding buildings and in particular St Paul's Cathedral from excessive damage, provision was made to remove up to 95 per cent of the polluting gas in a gas washing plant. Here the flue gases were passed through giant cast-iron washing chambers 106ft in height, 53ft wide and 48ft deep. They were packed with cedarwood scrubbers and sprayed with river water to which a 50 per cent slurry of chalk could be added.

Aerial view of Bankside Power Station in the 1960s.

Chemistry laboratory, Bankside Power Station, in the 1960s.

The effluent water, after it had done its job and removed all acidic gases, was de-aerated and any traces of oil removed before being returned to the river. The gases were then discharged to the atmosphere. To preserve his elegant design, Scott incorporated four chimneys within the one 320ft-high tower. Each was fitted with a nozzle at its outlet to enable the plume to be discharged to the atmosphere at an increased velocity. The 'smokeless shimmer of vapour' was indeed a noble and familiar site on the London skyline.

Bankside's heyday was in the sixties and early seventies, and there were many distinguished visitors: the Queen came in 1962, Prince Charles in 1968 and the Lord Mayor of London in 1975. In autumn 1970, output levels were broken twice: on 30 September, 5,876,246 units were sent out over a twenty-four-hour period; but on 8 October, this rose to 6,004,364 units! Then, in 1973, came the Arab–Israeli war, resulting in the oil crisis of the mid-seventies and an enormous increase in the price of oil. This was fatal for oil-fired Bankside, and the station never really recovered. It simply could not compete, in price terms, with larger coal-fired stations or nuclear power. The final outcome was inevitable!

2 Gas

Before natural gas was discovered in the North Sea, practically all gas used for commercial and domestic purposes was manufactured from coal.

It has been known for many years that when coal is heated in the absence of air, gas is produced. As long ago as 1690, John Clayton, a Wakefield clergyman, was experimenting with the distillation of coal and observed, 'first there came over only fleghm, afterwards a black oyle and then likewise a spirit arose'. He collected the 'spirit' in a pig's bladder, which he subsequently lit with a candle flame to produce a luminous jet. Clayton was, however, unaware of any practical application for this strange phenomenon. Later, in 1760, George Dixon filled a kettle with coal, heated it and lit the gas coming from the spout. Encouraged by the luminosity of the flame, Dixon built a pilot plant but an explosion halted any further progress. Later, gas was obtained as a by-product by the Earl of Dundonald. His main purpose was the formation of tar for caulking ships but – as a diversion – he lit his home, Culross Abbey, with coal gas. It was left to the Scotsman William Murdoch in the late eighteenth century to realise the enormous commercial potential of gas.

Murdoch was employed by Boulton and Watt, the famous Birmingham-based manufacturers of steam engines, and in 1777, the 25-year-old engineer was dispatched to Redruth in Cornwall to take responsibility for his employers' steam engines in that county's tin mines. Murdoch had an inventive mind – he is said to have filled the head of his pipe with coal and ignited the gas as it emerged from the stem. It was not long before he astonished his neighbours by lighting his house with coal gas. That was in 1792, and in 1798, back in Birmingham, he succeeded in lighting Boulton and Watt's Soho factory in that city. Quick to realise its commercial importance, Boulton and Watt were soon illuminating the Salford cotton mill of Phillips and Lee with gas. And money was saved! Previously, the mill owners had paid over £2,000 per year for candle illumination compared with £600 for lighting by the much brighter gas burners. At this stage, however, gas lighting was seen as suitable only for illuminating factories. But all was to change when an extrovert Moravian, Friedrich Albrecht Winzler, arrived in London.

Frederick Albert Winsor, as he now styled himself, set up home in Pall Mall in 1803. He had heard about gas lighting from the Frenchman Philippe Lebon, claimed it was his invention and set about publicising this new source of light with pamphlets and lectures. In his own words, 'the thought of introducing the discovery for the advantage of the British realm struck me like an electric shock'. To his credit, Winsor was first to deliver gas by pipes to street lamps. He was soon to gain the approval of the Prince Regent and on the evening of 4 June 1807, gave his first public demonstration of gas lighting to celebrate the royal birthday.

The carbonising furnaces were situated in Winsor's Pall Mall house and 1½in-diameter pipes led the gas to a series of lamps in the gardens of the Prince Regent's residence at Carlton House. Here they illuminated a board inscribed with the following ode:

Sing praise to that power celestial,
Whom wisdom and goodness adorn!
On this day — in regions terrestrial,
Great George, our lov'd Sov'reign, was born.
Rejoice, — rejoice, 'tis George's natal day.

Without question, Winsor's display for the Prince Regent was well received, *The Monthly Magazine* recording:

The light produced by these gas lamps was clear, bright and colourless, and from the success of this considerable experiment hopes may now be entertained that this long talked of mode of lighting our streets may at length be realised. The Mall continued crowded with spectators until nearly twelve o'clock and they seemed much amused and delighted by this novel exhibition.

Following his success, Winsor set about promoting the virtues of gas lighting at every opportunity. Not surprisingly, oil and candle sellers were keen to see the back of Winsor's ideas. They claimed that gas lighting would ruin the navy because seamen would no longer be able to be recruited from whaling ships if gas put an end to the demand for whale oil. As well as that, the benefits of gas lighting were frequently met with ridicule. People were unconvinced by the idea of 'lighting London with smoke' and 'carrying light below the streets in pipes'. Sir Walter Scott commented that 'there was a mad man proposing to light London with – what do you think? Why with smoke.' When the House of Commons was eventually lit by gas, members placed their hands on the pipe work and were amazed to find that it was not hot! The sceptics were even joined by members of the scientific community. Sir Humphry Davy remarked mockingly: 'Is it intended to take the dome of St Paul's as a gasholder?' Another eminent scientist observed that 'it would be as easy to bring down a bit of the moon to light London as to succeed in doing so with gas'. Public sentiment was neatly summed up in a contemporary rhyme:

> We thankful are that sun and moon
> Were placed so very high
> That no tempestuous hand might reach
> To tear them from the sky.
> Were it not so, we soon should find
> That some reforming ass
> Would straight propose to snuff them out,
> And light the world with gas.

But lighting the world with gas was precisely Winsor's aim. He even proposed that there should be a tax on lighting by candles or oil and estimated that gas lighting would save the country a massive £128,041,667.

Gas Light & Coke Co.

Winsor's ideas soon bore fruit, for in 1807, at the 'Crown & Anchor Tavern' in the Strand, the first meeting of the New Patriotic Imperial and National Light and Heat Co. was held. It was not long before this became the Gas Light & Coke Co. (GLCC). Then, in 1812, it was granted a Royal Charter by the Prince Regent, on behalf of George III, giving the company power to raise capital with limited liability, to dig up streets and lay mains to supply gas to the City, Westminster and Southwark. The company, by virtue of being established both by Royal Charter and Act of Parliament, had a governor and a court, rather than a chairman and board of directors. James Ludovic Grant became the company's first governor. Hopes were high. Up to now, London had been illuminated by tallow candles or oil lamps but the German chemist Frederick Accum, formerly assistant to the renowned Sir Humphry Davy at the Royal Institution, insisted that one gas lamp emitted light equal to three tallow candles or eighteen oil lamps. It was intended that Winsor would be the company's engineer. But the poor man was heavily in debt and refused to do any work unless the company paid him what he alleged they owed. A stand-off resulted and for the moment Winsor left the scene.

The company's first works were at Canon Row, Westminster, not a stone's throw from the Palace of Westminster, and their offices were at 96 Pall Mall. The company were full of ambition, with plans to light the approaches to Parliament to gain maximum publicity and to obtain a contract to light the Liberty of Norton Folgate in Spitalfields, just north of the City. The problem was that they overstretched themselves and failed to deliver. But the day was saved by the appointment of Samuel Clegg, who was paid a salary of £500 per year. (It was said of Clegg 100 years later that 'perhaps there is not another individual to whose zeal and ability the art of gas making is so much indebted for the variety as well as extensive utility of his inventions and improvements'.) He had previous experience working for Boulton and Watt with William Murdoch and realised at once that the Canon Row works were entirely inadequate. The outcome was a move to a much larger site at Providence Court, Great Peter Street (at the junction of Great Peter Street and the present-day Horseferry Road). So was born the Westminster Gas Works, the first permanent gasworks for public supply in the country. And it was not long before a further works was constructed in Curtain Road, Shoreditch, to fulfil the Norton Folgate contract.

The manufacture of gas in those early days at the Westminster works of the Gas Light & Coke Co. is described by George Dodd in his *Days at the Factories*, in which he tells of a visit he made in 1843. The town gas (as it was known) was separated from coal by distillation, in the absence of air, in iron retorts. To begin with, horizontal retorts were used but these could operate only by a batch system and so had to be frequently recharged. Later, vertical retorts were employed, which were continuously charged at the top with the coke which then fell to the bottom. The coke residue was then either sold or used to further heat the retorts. Many impurities were produced which had to be removed before the gas was sold. First of all the evolved gas was passed through pipes surrounded by cold water to condense out tars, then ammonia was removed by passing through water and finally hydrogen sulphide by passing through lime water. The town gas (including 55 per cent hydrogen, 30 per cent methane and 10 per cent carbon monoxide) was then stored over water in gas holders. Due to the public's fear of explosions, the first gas holders were enclosed within brick buildings. In fact the risk was minimal, as convincingly demonstrated by Clegg who put a lighted taper to one by way of demonstration. The traditional gas holder was a wrought-iron container, sealed in a tank of water. Gas was admitted at the base causing the container to rise. Its capacity could be increased by telescoping the inner container within a series of outer ones and the whole was surrounded by a trellis of wrought iron.

The Westminster gasworks operated from 1813 until 1875, and in its first year produced 20 million cubic feet of gas. To begin with there was just one gas holder with a capacity of 14,000 cubic feet, but by 1822 a total of eighteen had been installed, the first four holders being enclosed within brick buildings. In 1856, the works was visited by the Prince of Wales (the future Edward VII). The plant closed in 1875, when the giant Beckton gasworks opened (see later), but gas holders remained on the site until 1937. It was at this time, during the redevelopment of the site, that 6,000 gallons of tar were discovered, buried! In the Second World War special underground shelters were constructed in the plant's foundations to secure the safety of government officials and it is said that these were linked by an underground railway to other shelters in Whitehall.

In the early days of gas supply, explosions were an ever-present hazard. Barely a few weeks after the Westminster works opened, Clegg was foolish enough to allow a purifier to remain in operation after it had developed a fault. The result was a deafening explosion prompting the Home Secretary, Lord Sidmouth, to invite the Royal Society to investigate the safety of gas manufacture. Sir Joseph Banks and Sir William Congreve were conducted around the works by Clegg and, to their consternation, he punctured one of the gas holders and lit the escaping gas – there was no explosion. They were obviously convinced, commenting that the previous incident was only the equivalent of five to ten barrels of gunpowder. As it turned out, there was a beneficial side-effect to the Royal Society's visit. Sir William Congreve became a convert. He took the lead in arrangements to celebrate – prematurely as it turned out – the exile of Napoleon Bonaparte to the island of Elba. A large pagoda in St James's Park was to be illuminated by gas light in the presence of the Allied sovereigns and other dignitaries. Unfortunately, Congreve opened the fete by organising a fireworks display, a spark from which burnt the pagoda to a cinder. Even so, the company came out of it well and soon gas light was installed on a permanent basis in St James's Park.

Explosions were not the only hazard – there was pollution as well. The Commission of Sewers refused to accept effluent and tar from the Westminster works. The company's solution was simply to store it in tanks and then pump it directly to the Thames – and this at a time when Londoners were still taking their drinking water from the tidal river.

Meanwhile, as a result of repeated calls for compensation, Winsor was finally paid an annuity. He bought shares in the company and took his seat on the Court. It was not for long, for in 1815 he fled the country (and his creditors). He died in Paris in 1830. Concurrently, the GLCC purchased the premises of the Golden Lane Brewery in Brick Lane, Old Street, and built a works there to supply the area from Goswell Road to Holborn.

Other Gas Manufacturers

Other manufacturers were soon to follow the lead of the GLCC. One such was the City of London Gas Light & Coke Co., founded by William Knight. Knight, who had previously been prosecuted 'for making of gas light and making divers large fires of coal and other things whereby noisome and offensive stinks and smells and vapours resulted', set himself up in Dorset Street, Blackfriars, in premises previously occupied by the New River Company and supplied the Fleet Street and Ludgate Hill area.

Gas supply south of the river began at Bankside. In 1814, the Bankside Gas Works came to the area. It was constructed by Munroe & Co. on land previously occupied by a glasshouse belonging to the glass maker Stephen Hall. The works were just west of Pike Gardens and on the present site of Tate Modern. It had one gas holder, a retort house, various sheds, office accommodation and a house for the manager.

In 1828, the Bankside Gas Works was taken over by the Phoenix Gas Light & Coke Co. and became known as the Phoenix Gas Works. It made good use of its site by the Thames; scrubbers used river water to remove tar and ammoniacal liquor from the heated coal. These waste products were collected in barges lying alongside the works and taken away for disposal. Many eminent men were associated with the enterprise, including Lord Holland, nephew of Charles James Fox; James Abercromby, Speaker of the House of Commons; Thomas Fowell Buxton, brewer, social reformer, MP and successor to William Wilberforce as head of the anti-slavery party; and the brewers Barclay Perkins. Board meetings were usually held at the works, or else at the 'Three Tuns Tavern' off Borough High Street.

Throughout the nineteenth century there were many improvements to equipment. In 1861, new steam-powered grabs were installed to ease the removal of coal from the Thames-side barges and, in 1866, £3,521 was spent on a new coal store.

The company had many well-known customers. In 1824, a deputation was received at one of the company's board meetings from St Olave's parish church. It was agreed that the church should be lit by gas lights supplied from a meter and a price of 15s per thousand cubic feet was agreed. In 1837, the nearby Barclay Perkins brewery agreed a similar arrangement. However, relations with customers were not always harmonious. The Phoenix supplied gas to light London Bridge and there was a long-standing run-in with the Bridge House Estates Committee, the authority which financed and maintained the City of London's bridges. In 1836, the Committee complained about broken glass in the gas lights and also that the lamp lighter, whose job it was to light up the bridge, was arriving late for this important task. The Phoenix responded simply by changing the lamp lighter. More trouble arose in 1841 when the Phoenix got fed up with the exposed lamps on the bridge becoming repeatedly damaged. They went so far as to threaten to cut off the supply.

More unusual customers were the many intrepid balloonists. Ever since the Montgolfier brothers flew over Paris at the end of the eighteenth century, ballooning had become a fashionable amusement. In 1824, the company were asked to inflate a balloon for a Mr Smithers who wished to ascend from Montpelier Gardens in Walworth. He later changed his mind and chose the 'Green Man' pub in the (Old) Kent Road. The company were happy with this arrangement 'provided there was sufficient security for payment'. There was also a Mr Green who was involved with the 'diversions' at Cremorne Gardens, by the river at Chelsea. He bought gas for balloon ascents from the Vauxhall works of the London Gas Co. Green ascended accompanied by a lady and a leopard! There were many other ascents from Cremorne Gardens. In 1864, 'the flying man', Eugene Godard, was unfortunate enough to come down in his balloon, *The Czar*, from 5,000ft onto the spire of St Luke's Church, Chelsea, with fatal results. It was not long before the gardens were closed for being 'the nursery of every kind of vice'.

Phoenix Gas Works, Bankside, Southwark, in the nineteenth century.

In time, the South Metropolitan Gas Co. became the dominant force in south London. The company was founded in 1829 and appointed George Holsworthy Palmer to build their first works on a 3-acre rural site, adjoining the Grand Surrey Canal on the eastern side of the Old Kent Road. The works began operation in 1833. In common with other gas undertakings, the South Metropolitan was not immune from explosions. In October 1836, a workman with a light walked into the purifying house unaware that there was an escape of gas. The resulting explosion was 'felt all over London' and the building was destroyed, injuring the workman in the process. His doctor's bill came to £18, which the company reluctantly paid 'after careful consideration'.

In 1839, the South Metropolitan appointed Thomas Livesey, then working at Brick Lane for the Gas Light & Coke Co., as chief clerk. So began a seventy-year association of the Livesey family with the South Met. Livesey came with much experience, and he and his descendants were to shape the company's course throughout the nineteenth century. In 1842, Livesey introduced a workmen's sick benefit fund, followed by a superannuation fund in 1855.

The Phoenix were in bitter competition with their rival, the South Metropolitan. Chaos resulted in the streets of south London with both companies laying pipes in the same street and competing for business from the same householders. At one time the Phoenix were selling their product at 9s per thousand cubic feet, with the South Metropolitan charging 11s. 'The purity and illuminating power of our gas is such that 3,000 cubic feet will burn as long and give as great a light as 5,000 cubic feet of the gas of the Phoenix,' retorted the South Met. But the Phoenix won the day and the Old Kent Road company were forced to reduce their prices. As time passed, the two companies agreed to have 'spheres of influence' and in 1853 a deal known as the 'district agreement', whereby only one company would have mains in the same road, was introduced. In 1860, the Metropolitan Gas Act eliminated much unnecessary competition by giving companies a monopoly of supply in their particular area and also established a quality standard – the Twelve Candle Standard. The standard referred to the illuminating power of a quantifiable open gas flame. Gas referees were appointed to ensure that companies complied with the Act.

The South Metropolitan expanded rapidly. In the 1860s, a church near their works was demolished to make way for extra capacity. The company, however, contributed £5,000 towards the cost of a new church and made land available. It was at this time that the South Metropolitan employed many Irishmen and because of Fenian unrest they were all required to sign an address expressing 'loyalty to the throne and disapproval of the acts of the Fenians'.

Thomas Livesey, who died in 1871, was succeeded by his son, George. Then, in 1889, the Gas Workers Union was formed. Its leader was Will Thorne who soon negotiated a three by eight-hour shift system to replace the previous two by twelve-hour shift. But much to the displeasure of Will Thorne, George Livesey set up a profit-sharing scheme which allowed the men to become shareholders.

Expansion continued in the 1880s, by which time the South Metropolitan occupied 36 acres (compared with the original 3 acres), but space was still at a premium, prompting the company to acquire a further 130 acres at East Greenwich and soon to open a works there. Other works were to follow at Woolwich and by 1920 the company employed 9,000 people. George Livesey was rewarded with a knighthood by Edward VII for his services to the gas industry.

A visual reminder of the gas industry in south London is a stained-glass memorial window directly over the altar of the retro-choir (Lady Chapel) in Southwark Cathedral. It is dedicated to 386 employees of the South Metropolitan Gas Co. who lost their lives in the First World War. The centre light shows the Presentation of Our Lord in the Temple and those on either side the Three Kings and Simeon. The window, produced by John Hardman & Co. of Birmingham to designs by Donald Taunton, was unveiled with great ceremony on 8 December 1923 after sunset and so measures had to be taken to illuminate it. Appropriately, it was lit from outside by a series of gas burners, the light from which was reflected onto the window by a white screen to produce a diffuse light 'almost indistinguishable from sunlight'.

Meanwhile, much was happening north of the river. The GLCC were the dominant force, but by 1823 they were joined by the Imperial Gas Light & Coke Co. with works in Shoreditch, and the City of London Gas Light & Coke Co. at Blackfriars. Competition was not always fair. On one occasion men from the Equitable Company (with works in Pimlico) inserted a valve into the GLCC's 6in main and partially closed it, thereby exacerbating the discontent of their rival's customers who now got their gas at reduced pressure. And that was not all – they even put lumps of clay into the mains.

Preserved gas holder No 2 of the Imperial Gas Light & Coke Co., Fulham, 1830 – the oldest gas holder in the world.

The City company, operating in Blackfriars, were renowned for their harsh treatment of their employees. Thomas Sheppard was foolish enough to steal some pipes and ended up being publicly whipped and serving six months' hard labour in Clerkenwell's House of Correction. Mark Pember, a lamp lighter, was clapped in Newgate Prison for borrowing a spade to do a job at his home. And woe betide any of the company's customers who got behind in settling their bills – they could find themselves in the King's Bench Prison, Southwark, for debt.

The Imperial Gas Light & Coke Co. became a statutory concern in 1821. They were soon the largest gas company in London, their area including Pimlico, Chelsea, Kensington, Knightsbridge, Marylebone, St Pancras, Mile End and Whitechapel. The Imperial had works at St Pancras, at Shoreditch on the Regent's Canal, and later at Fulham.

By 1860, eight companies were operating north of the river. These were the Gas Light & Coke Co. (Westminster, Brick Lane and Curtain Road), the Imperial (Shoreditch, St Pancras and Fulham), the City of London (Blackfriars), the Independent (Haggerston), the London (Vauxhall), the Equitable (Pimlico), the Western (Kensal Green) and the Great Central (Bow Common). South of the river were the Phoenix, the South Metropolitan and the Surrey Consumers.

Explosions at gasworks were far from rare. The *Illustrated London News* of 4 November 1865 reported how a gasometer of the London Gas Light Co. at Nine Elms exploded killing several people on the spot. Nine men later died and many were disfigured. The fault lay with something described as a 'governor' which had been stepped on by accident. Following an inquiry it was recommended that governors should in future be covered or protected so that persons unacquainted with their nature should not have access to them.

Beckton

In the years that followed, many companies consolidated and merged and of these the Gas Light & Coke Co. dominated. In 1848, they appointed Simon Adams Beck to their board; he became deputy governor in 1852, and governor in 1860. It was Beck who, on 19 November 1868, drove the first pile into the river wall at Galleons Reach, near Barking. On the following day the Court of the GLCC resolved that the company's new property and works at Barking Creek should henceforth be known as Beckton. The Beckton works covered some 627 acres and became the largest gasworks in the world. Today, although the works are long gone, the area is still known as Beckton.

Up until the 1880s, gas was used exclusively for lighting, but then it began to face competition from the newly commercialised electricity. A new outlet was found in cooking and the Gas Light & Coke Co. responded to the opportunity by opening the first gas showrooms.

The GLCC continued to expand by taking over other concerns; by the late 1880s had absorbed the City of London, the Great Central, the Equitable, the Western, the Imperial, the Independent and the Victoria Docks Co. at Silvertown. By 1934, it had works in Shoreditch, Kensal Green, Fulham, Bow Common, Beckton, Nine Elms, Bromley, Stratford, Harrow, Brentford, Southall, Staines and Southend.

While the broad principles of gas manufacture were the same as those described by Dodd in his *Days at the Factories*, many refinements were made in later years. In the early days, gas (as bubbles) was passed through a solvent liquid to remove impurities. Later, scrubbers were used where a counter-flow of gas passed over a film of solvent soaked onto a solid support. Rotary washers were introduced later. In these, wheels are rotated – with balls or brushes attached to them – through solvent. The gas is then passed over the wetted surface. For the removal of hydrogen sulphide, passage over iron oxide proved superior to lime washing. Naphthalene, which has a tendency to form deposits on solid surfaces, was removed by oil washing from the early twentieth century. Until the late nineteenth century the horizontal retort was the sole means of carbonisation but was replaced by arrays of vertical or inclined retorts where coal was charged from the top, carbonisation taking place as the coal moved slowly down.

Gas holders still remain a familiar feature of the urban landscape. They serve to balance a fluctuating consumer demand against the uniform rate of gas manufacture, to store gas, and, by virtue of the weight of the bell, to force the gas through the distribution mains. Waterless gas holders were invented in Germany in 1916. They contain a piston which rises and falls as needed.

At the beginning of the Second World War the GLCC were by far the largest gas manufacturers in the country, supplying almost 15 per cent of the entire UK output. Not surprisingly, they fell victim to air raids. A total of 200 bombs fell on Beckton, and on 'Black Saturday', 7 September 1940, wave after wave of German aircraft bombarded London, and Beckton, Bow Common, Bromley-by-Bow and Stratford were all put out of operation.

Horizontal retorts, Beckton gasworks. *(By permission of Newham Local Studies Library)*

Beckton was the jewel in the crown of the Gas Light & Coke Co. Its opposite number and near neighbour on the other side of the river was the South
Metropolitan's East Greenwich gasworks. In its heyday seventeen colliers delivered as much as 1 million tons of coal every year from the Durham coalfields
to purpose-built jetties at Beckton. The company also took advantage of its by-
products of coal tar and ammonia by forming the Beckton Products Works in
1879 to produce road tar, sulphuric acid and many organic chemicals.

By the end of the war, deficiencies in the industry as a whole were becoming
ever more marked. The fundamental problem nationwide was the fragmentation
of and variations within the industry. In response, the Heyworth Committee,
reporting in 1945, recommended nationalisation. It was left, therefore, to Hugh
Gaitskell, Minister of Fuel and Power, to steer a bill through Parliament to bring
the industry into public ownership. Throughout the country, 1,064 local gas
undertakings were vested in twelve area gas boards, those works of the Gas Light
& Coke Co. going to the North Thames Gas Board.

Natural gas was well established as a fuel in the USA, but, as a report of
1952 states:

Although there are no appreciable known reserves of natural gas in Britain,
a discovery of any magnitude would be of immense benefit to our national
economy and it is suggested that prospecting, where there is any possible hope
of success, should continue to receive vigorous support.

Beckton gasworks in the 1960s. *(By permission of Newham Local Studies Library)*

And prospecting was fruitful, for on Friday 17 September 1965, the *Sea Gem*, operated by BP, discovered gas 48 miles off Grimsby – it soon yielded 10 million cubic feet of gas every day.

Today, natural gas has replaced coal gas but because of its different chemical composition all appliances had to be changed. This was accomplished in a ten-year programme from 1967 to 1977. In 1986, the industry was privatised once more with the creation of British Gas plc, and other restructuring has taken place since.

3 Post & Telecommunications

The Postal System

In medieval England letters were conveyed from place to place by messengers. Plainly this was a preserve of the well-to-do; a letter would be handed to the messenger who would then mount his horse and deliver it. Most letters were sent by the King, who had many messengers in his service.

A national postal service emerged in the reign of Henry VIII, when the King's messengers began handling mail from the general public. In 1516, Sir Brian Tuke was appointed Master of the Posts, and so began the service of postal deliveries along important roads from London. At various stages along the route postmasters were appointed. They were often innkeepers at coaching inns and were required to have 'three good strong leather bags lined with baize or cotton and three horns to blow'.

Thomas Witherings became Postmaster General in the seventeenth century. He had studied postal services on the Continent and was instrumental in extending the service in this country. In 1635, a government proclamation decreed that a service should run day and night between Edinburgh and London and that mail should be delivered there and back within six days. The route was the Great North Road; by-posts were set up along the way to take and receive letters from important towns off the route such as Hull and Lincoln. Other services soon followed: for example, the route to Dover and the Continent had stages at Dartford, Gravesend, Rochester, Sittingbourne and Canterbury. There were similar services along Watling Street to Chester, thence to Holyhead and Ireland, and along the Bath Road to Bristol and the West. Payment depended on the number of sheets carried and the distance travelled: 2d for a single sheet going less than 80 miles, and 8d to Scotland. It was Thomas Witherings' proud boast that 'any man may with safetie and securitie send letters to any part of the kingdom and receive an answer within 5 days'.

To counter the rise in unofficial posts an ordinance was issued in 1654 decreeing that 'The Office of Postage of Letters Inland and Foreign [is designated] as the only carrier of letters in all places of England, Scotland and Ireland and to and from all other places within the Dominions of this Commonwealth', and postmasters on the route were to keep 'four good horses or mares at the least for the said post service'. Also, 'a stamp is invented that is put on every letter shewing the day of the moneth that every letter comes to the office, so that no letter carryer may dare to detayne a letter from post to post, which before was usual'.

All these measures increased efficiency. Witherings watched over his subpostmasters with an eagle eye, writing forcibly to them if any transgressions came to his attention. The postmaster at Faringdon was once on the receiving end: 'I am credibly informed that your horse tired at Lechlade last week and was there supplied by one out of a cart. This is scandalous.'

The Post Office's headquarters were in Bishopsgate. A comptroller, accountant and treasurer worked there, assisted by eight clerks, three window-men, three sorters and thirty-two letter carriers. It moved to Lombard Street in 1678 and to St Martin's Le Grand in 1829.

William Dockwra, a London merchant, lived in Lime Street, and it was here in 1680 that he set up his penny post. It was a significant innovation. For a standard fee of one penny, letters were delivered and collected within London and its suburbs. The head office was at Dockwra's house in Lime Street and he had offices for sending, receiving and sorting mail at Newgate Street, Chancery Lane, St Martin's Lane, Southwark and East Smithfield. Opposition was fierce – not only from the Post Office, which saw Dockwra's enterprise as a competitive threat, but also from porters, who vandalised posters advertising the service. That well-known firebrand Titus Oates called it a 'Popish Contrivance' and the 'work of the Jesuits'. The unfortunate Dockwra was prosecuted in 1682 and was forced to call a halt. It came as no surprise – bearing in mind its popularity – that his service soon reopened under the auspices of the official Post Office. It was known

GENERAL POST OFFICE
& ST MARTINS LE GRAND,
LONDON.

General Post Office, St Martin's Le Grand.

throughout the eighteenth century as the London District Post. Dockwra, however, did not go unrewarded: he was awarded a pension of £500 per annum 'in consideration of his good service in inventing and setting up the business of the penny post'. Daniel Defoe was similarly impressed, writing of 'this modern contrivance of a private person, one Mr William Dockwra, now made a branch of the General Revenue of the Post Office to send letters with utmost safety and dispatch – we see nothing of this in Paris, at Amsterdam, at Hamburg or any other city that ever I have seen or heard of'. By 1702, the London District Post was handling more than 1 million letters every year. The charge remained at 1*d* until 1794, when the penny post was expanded under the guidance of a former letter carrier, Edward Johnson. The number of carriers (modern-day postmen) in the town was increased from forty-four to 126, with six deliveries per day. In the suburbs (or 'off the stones', as they were known) the number increased from thirteen to ninety-one. In 1794, all deliveries were charged at 2*d* and then, in 1805, those to the suburbs increased to 3*d*.

Rowland Hill was born in Kidderminster in 1795. He had a successful early career as a teacher at a progressive school he had founded and was admired by no less a person than Jeremy Bentham. He later became secretary of the South Australia Colonisation Commission, an organisation to settle people in Adelaide. But Hill is remembered today as the founder of the Uniform Penny Post. It is said that Hill's interest in the post was aroused when he witnessed an impoverished young woman refusing to accept a letter from her fiancé. Her difficulty was that she (the recipient) was required to pay. If the recipient could not afford it, they simply refused to accept the letter. Due to the chaos this state of affairs produced, Hill was prompted to write his pamphlet *Post Office Reform: its Importance and*

Practicability in which he called for a 'low and uniform rate' based on weight and not distance. He worked out that the majority of the post's costs were incurred not in transport but in the laborious process of handling and sorting. He proposed that the sender pay with the now familiar adhesive stamp. Despite the anticipated objections to his 'wild and visionary schemes', Hill's plans were put into action and in 1840 rates were reduced to 1*d*. So was born the world's first adhesive postage stamp – the Penny Black.

Soon, letter boxes proliferated throughout the land. They were first suggested by Rowland Hill himself, although others say it was the author Anthony Trollope, who worked as a Post Office surveyor. London got its first letter boxes in 1855. Six were placed 'along the thoroughfare of Ludgate Hill, Fleet Street, the Strand, and Piccadilly ... on the side of the footway in such a position not to obstruct the traffic of any kind'.

But if messages needed to be sent and received quickly, an alternative to the simple letter had to be found. This proved to be the telegraph.

Telegraphs

The first telegraphs were conveyed by optical means. For instance, the Admiralty sent messages to the dockyards at Portsmouth by building a series of towers along the way and using semaphore signals to communicate one with the next. A message could reach Portsmouth (70 miles from London) in fifteen minutes. Then in the mid-1840s Samuel Morse in the USA and Cooke and Wheatstone in the UK invented the electrical telegraph.

The first electrical telegraphs were operated by private companies. One was set up by Sydney and Alfred Waterlow. Wires were strung between metal masts and over London's rooftops. Signals were sent by Morse code. There were many objections: for example, the Drapers' Company complained about the wires passing over their hall, insisting that they had 'the freehold from the centre of the earth to the canopy of heaven', but relented when offered 2*s* 6*d* rent. Others objected that 'the magnified linen posts and clothes lines render hideous our most public and best constructed streets dwarfing the apparent altitude of some of our finest architectural elevations'.

The London District Telegraph Co. was registered in 1859 and the Universal Private Telegraph Co. one year later. The District Co. intended to have 100 offices within 4 miles of Charing Cross. Each office would take messages and then send them to the nearest office to the recipient. Offices were located at railway stations, and the travelling public were alarmed to see what must have appeared to them as equipment 'out of this world'. It was compared to 'a mixture of the beer engine and the eight day clock, with strangely marked dials resembling the differential calculus framed in a gooseberry tart'. And there were accidents. John Durham recounts:

At about 11 o'clock one of the Islington omnibuses was proceeding over Tower Bridge, when suddenly a loud noise was heard and a gentleman was hurled from the top of the omnibus to the roadway, falling with great force on his head. It was found that the telegraph wire had snapped with the fury of the gale and coiled round the unfortunate man, dragging him from his seat. We are sorry to observe that this accident ultimately proved fatal.

But the days of private telegraph companies were numbered. The Post Office saw it as their right to send messages, even if they were sent along wires. So, in 1869, an Act was passed in Parliament 'to enable the Postmaster General to acquire, work and maintain electric telegraphs'. Henceforth, all telegraph messages were sent by the Post Office.

Barely ten years after the Telegraph Act of 1869, the Post Office faced another threat. It came from Alexander Graham Bell, who arrived in England in 1877 with his new invention, the telephone. One year later, the first telephone exchange, operated by the Telephone Co. Ltd, was set up at 36 Coleman Street. There were other private companies – the National Telephone Co. and the Edison Telephone Co. (one of whose managers was George Bernard Shaw) – but it was not long before they were absorbed by the Post Office, and by 1911 the last remaining private company was taken over.

Cable Companies

The telegraphic transfer of messages required reliable cables and Henley's, the firm founded by William Thomas Henley, provided this service. Henley was born in 1814 in Midhurst, son of a fellmonger and glover. He arrived in London as a young man of 16 and worked as a labourer for a silk merchant in Cheapside before becoming a docker at the newly opened St Katharine Dock. But science and technology were his abiding interests, in particular electromagnetism, and he soon persuaded the owner of a chemist's shop in Commercial Road to allow him to exhibit and sell his home-made electrical machines. He was also fortunate enough to make the acquaintance of Professors Daniell and Wheatstone, both of whom were at King's College. By 1837, Henley had set up his own establishment, first at Red Lion Street, Whitechapel, and then in Haydon Street, Minories, making insulated conductors and electrical apparatus. Electrical telegraphs were established in the mid-nineteenth century and Henley was well placed to exploit this growing market. Expansion was rapid, leading to a move to 46 St John Street, Clerkenwell, where he employed twenty-three people, and then to Enderby's Wharf in East Greenwich.

Henley's specialised in the manufacture of long lengths of telegraph cable. They linked London with Manchester, Liverpool with Manchester, and Dublin with Belfast. The firm later expanded to make submarine cables and soon built new riverside works at North Woolwich.

Standard Telephone & Cable Co.'s Riverside Wharf. *(By permission of Newham Local Studies Library)*

William Henley (d. 1882), can be thought of as a founder of the modern communications industry. The firm he founded also made house wiring, telephone cables, power cables for the electricity supply industry and much more. In 1959, Henley's were acquired by AEI, then GEC, and are now part of TT Electronics.

There were other manufacturers. S.W. Silver started out in Greenwich in the early nineteenth century, and then moved across the river to just west of North Woolwich. The area, Silvertown, took its name from him. Charles Hancock (formerly of the Gutta Percha Co.) joined him and they set about making and laying cables. The firm became known as the India Rubber, Gutta Percha & Telegraph Works Co. Copper wire was used as the conducting medium. First it was wound onto bobbins and then several lengths twisted into a wire rope which was insulated by gutta percha. The cable was coated with jute and was further protected by windings of iron wire.

The original Gutta Percha Co. had been founded in 1845 by Charles Macintosh and Henry Bewley. They made the insulated core used for the 1850/51 cross-channel cables. As mentioned above, Charles Hancock subsequently joined S.W. Silver, following a dispute. In 1864, the Gutta Percha Co. merged with Glass, Elliot & Co. to form the Telegraph Construction & Maintenance Co. (Glass, Elliot shared a site at Enderby's Wharf in Greenwich with Henley's until the latter – because of obvious difficulties – moved north of the river.)

The Telegraph Construction & Maintenance Co.'s first contract was for a cable to cross the Atlantic. It failed, so they did not get paid. In 1866, they tried again,

COVERING THE CORE WITH TAPE

WIRING A "SHORE-END"

FINISHING

SHIPPING CABLE "S.S. SCOTIA"

THE PAYING-OUT MACHINERY

PICKING-UP MACHINERY

A VISIT TO THE TELEGRAPH CONSTRUCTION AND MAINTENANCE COMPANY'S WORKS, EAST GREENWICH

Telegraph Construction & Maintenance Co., East Greenwich, 1871.

with the *Great Eastern* as the cable-laying ship, and this time were more successful. More cable was laid by the *Great Eastern* in 1869, 1873 and 1874. The company was acquired by British Insulated Callender's Cables (BICC) in 1959.

Johnson & Phillips, founded in 1875 and located in Charlton, offered 'complete equipments of cable machinery, accessories and stores for cable laying and repairing steamers'. They also made telegraph, telephone and insulated cables. Johnson was born in Jersey, and worked for a while in Lancashire before joining the Telegraph Construction & Maintenance Co. as a draughtsman. He was later chief engineer responsible for laying the British Australasian cable. S.E. Phillips had worked for many years at Henley's. In 1875, they joined forces at the Victoria Works in Charlton.

Hooper's Telegraph Works were alongside Millwall Dock on the Isle of Dogs, and Siemens were at Woolwich. Siemens began in Woolwich in 1863 and, using the ship *Faraday*, laid the first transatlantic link between Ireland and the USA in 1870.

4 Water Supply & Sewage Disposal

Water Supply

London owes its existence to the River Thames and its earliest inhabitants took their water supply from the river and its tributaries. These sources were perfectly adequate throughout the early Middle Ages. William Fitzstephen, in the preface to his biography of Thomas Becket written in about 1180, tells us that 'there are also round about London in the suburbs most excellent wells whose waters are sweet, wholesome and clear. Among these, Holywell, Clerkenwell and St Clements are most famous …'

Conduits

In time, London's water supply was supplemented by that from conduits. (The term 'conduit' is misleading and could mean a pipe or a trench, or the point where the supply is drawn off.) In 1237, the first conduit, the Great Conduit, took water from the River Tyburn on land belonging to Gilbert de Sanford and conveyed it, by the force of gravity, to Cheapside in the City. The pure spring water, in earthenware or lead pipes, was conveyed from Sanford's land – where Stratford Place now stands, near Bond Street Underground Station – via Constitution Hill, Trafalgar Square, and over the Fleet River to Cheapside, a distance of about 3 miles. At Cheapside it was the job of water carriers, or 'Cobs' as they were familiarly known, to bring water in wooden buckets slung over their shoulders to people's houses.

Private citizens were free to collect water for themselves if they so wished. A charge was made, however, for certain tradesmen, which went towards the upkeep of the conduit. In 1312, for instance, citizens were appointed to 'faithfully collect the money assessed upon brewers, cooks, and fishmongers at their discretion for the easement they enjoy of the water of the conduit of Chepe, and to expend the same upon the repairs and maintenance of the said conduit'.

In 1440, further plans were made for extending London's water supply. The Abbot of Westminster Abbey gave permission for a conduit to be laid from springs in the manor of Paddington for a yearly cost of 2lb of pepper payable on

Water carrier. *(By permission of Thames Water)*

Below: Gathering water at the conduit. *(By permission of Thames Water)*

the feast of St Peter ad Vincula. And as London's population increased, so did its demand for water. In 1535, water was taken from springs at Dalston and conveyed to Aldgate. Then, in 1544, an Act of Parliament, the London Conduit Act, enabled water to be conveyed from 'dyvers great and plentifull springes at Hamsptede Hethe, Marylebone, Hakkney, Muswell Hill and dyvers places within five miles of the said Citie'. John Stow tells us, in his Survey of 1603, of further conduits at Aldermanbury, Fleet Street, Grass Street, Stocks Market (near to the present Mansion House), Bishopsgate, London Wall and Lothbury. Present-day street names give us a clue to their location – for instance, Conduit Street and Lamb's Conduit Street.

Broken Wharf Water Works

By the sixteenth century London's population had grown considerably and in consequence so had its need for water. Fortuitously there had been advances in pumping technology, and it was not long before pumps were used to raise water to a head and so feed it by gravity to its point of consumption. Pumps would be worked by muscle power – usually a horse gin. An early water works based on this principle was at Broken Wharf, in Upper Thames Street, downstream of Blackfriars. Stow tells us that 'one other new forcier [pump] was made neare to Broken Wharfe to convey Thames water into men's homes of West Cheape, about Powles, Fleet Street &c by an English Gentleman, named Bevis Bulmer in the yeare 1594'. Bevis Bulmer was a mining engineer. He used a chain pump driven by horses to raise river water to about 120ft in a wooden tower. The tower can be seen in Hollar's View of London, 1647.

Somerset House Water Works

Another horse-powered water wheel was located within the precincts of the original Somerset House, near to the present Temple Underground Station; nearby Water Street is a reminder for us. It was built in about 1655 by Sir Edward Ford, who pumped water to a wooden pyramidal tower, 120ft in height. Somerset House was soon to be occupied by Henrietta Maria, wife of Charles I, and unfortunately for Ford and all his customers, she objected to being overlooked by the tower. At the Restoration, her son, Charles II, came to her aid and promptly ordered that 'the great fabric of wood which they have erected for raising water from the Thames, on the soil of the river, is a nuisance and must be removed within three months'. And that was the end of the Somerset House Water Works.

London Bridge Water Works

It was the Dutch (or maybe German) engineer Peter Morris, patronised by the Lord Chancellor, Sir Christopher Hatton, who opened the London Bridge Water Works in 1582. He relied upon the restricted flow of water as it passed beneath the bridge. The bridge, built in the late twelfth century by Peter of Colechurch, had twenty-five arches, with starlings (protective coverings) on the piers. These had the effect of reducing the flow of the Thames and provided the means for

Morris to get a head of water with a water wheel which he constructed in the first arch of the bridge. A second pump was added one year later. Water was conveyed to individual houses by pipes, for which Morris made a charge. The remaining water passed to a standard where all could obtain a free supply. Stow takes up the story thus:

> A certain German named Peter Morris, having made an artificial forcier conveyed Thames water in pipes of leade at the north end of London Bridge and from thence into diverse mens houses in Thames Street, New Fish Street, up to the northwest corner of Leaden Hall, the highest ground of all the Citie, where the waste of the maine pipe rising into this standarde (provided at the charges of the Citie) with foure spoutes did at every tide runne foure wayes, plentifully serving to the commoditie of the inhabitants neare adioyning in their houses.

Morris first demonstrated his idea by directing a jet of water over the steeple of the nearby church of St Magnus the Martyr. So impressed were the City Fathers that they granted him a lease of the first two arches of the bridge for 500 years for his water wheels. Morris' works consisted of an undershot water wheel which, via a crank and connecting rod, drove pump barrels of 'brasse or iron unto each of [which] must be fitted a forcer well leathered'. The forcer was a plunger lined with leather. The Morris family retained their interest in London Bridge Water Works until the lease was transferred to Richard Soame, a City goldsmith in 1701, who at once employed George Sorocold, an experienced water engineer, to modernise the works. Soame also acquired the Broken Wharf works. By 1702,

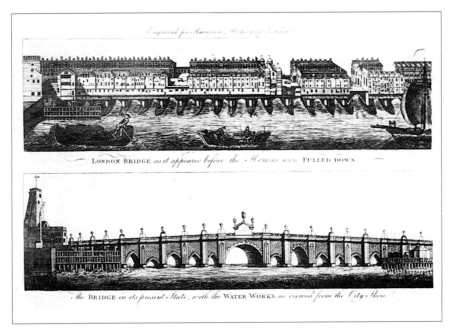

London Bridge, with the water works on the left. *(By permission of Thames Water)*

a fourth arch was occupied and water pumped as far away as 'Goodman's Fields, Hounsditch and White Chapel'. A full description of the works was given by Henry Beighton FRS in 1737, and is reproduced in H.W. Dickinson's *Water Supply of Greater London*, in which it is estimated that over 107,000 gallons of water were raised to a height of 120ft every hour.

In 1767, John Smeaton designed a plant to serve Southwark, south of the river, where he installed a water wheel in the second arch from the Southwark end of the bridge.

At times, particularly in hot summers and at low tides, the water wheels proved insufficient to serve the demand, and so a Newcomen atmospheric engine was incorporated to supplement supply. However, by the early nineteenth century, the days of the London Bridge Water Works were numbered. The rival New River Company was able to supply water at a greater height, and it was also purer. The end came in the 1820s when old London Bridge was finally demolished to make way for John Rennie's new structure.

The New River Company

By the sixteenth century London's population had increased to about 200,000. To serve this ever-increasing number Edmund Colthurst, in about 1600, proposed that water should be taken from the springs at Amwell and Chadwell in Hertfordshire and fed to London. So began the embryonic idea of the New River Company. James I granted Colthurst letters patent in 1604, but the City Fathers were worried about his financial security, so, in 1609, they accepted an offer from Hugh Myddelton, a London goldsmith, and obviously a man of means. Myddelton was born in North Wales and had interests in the clothmaking industry, and in lead and silver mining. He began his scheme at Chadwell, but progress was not always unhindered. There were objections from landowners, forcing the City Fathers to appeal to the Privy Council to 'mediate to the king's most excellent majesty that all lette and hindraunce may be removed and way given to Mr Midleton'. James I was obviously impressed, for he ended up giving his own money to the project and permitted Myddelton to bring the channel through his estate at Theobalds. The King provided half the cost, for which he was entitled to half the profits. The monarch continued to benefit from the New River until Charles I gave up the royal interest for a yearly payment of £500 – a sum known as the Crown Clog.

After the royal intervention, progress was unimpeded and the New River continued its course, sometimes traversing valleys in lead troughs on elevated timber frames, to eventually reach New River Head in Islington. Its total length was about 40 miles, with a fall in level along its route of only 18ft. Stow describes it in contemporary terms:

> The depth of the trench [in some places] descended full thirty foot, if not more; whereas [in other places] it required as sprightful Art againe to mount over a valley in a trough, between a couple of hills and the trough all the while born

up by the wooden arches; some of them fixed in the ground very deep, and rising in height above twenty three foot.

These elevated troughs were at Highbury and Bush Hill, Edmonton – that at Bush Hill known as the Boarded River.

From the beginning, the threat of pollution was realised and an indenture of 1612 stated:

> Wee comaund all pson and psons whatsoever that they or anie of them doe not hereafter cast or putt into said new river anie earth rubbish soyle gravell stones dogs catts or anie cattle carrion or anie unwholesome or uncleane thing nor shall wash nor cleanse anie clothes wool or other thinge in said river.

As well as by James I, the costs of the scheme were borne by Myddelton and twenty-eight 'Adventurers'. The New River was incorporated by Royal Charter in 1619; its seal showed rain flowing from an open hand onto the City of London with the inscription 'Et plui super unam civitatem' – 'And I rained upon one city'. The river terminated in Islington 'at an open idell poole', which became known as the Round Pond. It was 82ft above the level of the Thames and about 210ft in diameter.

At New River Head, near present-day Rosebery Avenue, water was conveyed to Londoners' homes in pipes made from hollowed-out elm trees. These were made by, amongst others, the carpenter Alexander Hay and his sons, John and Joseph, at Pipe Borer's Wharf in Southwark. Pipe Borer's Wharf later became the famous Hay's Wharf, 'larder of London', and is now the modern-day Hay's Galleria.

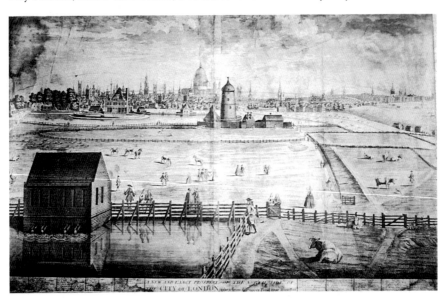

View of London with the New River, Islington, in the foreground. *(By permission of Thames Water)*

The New River was formally opened by the Lord Mayor of London in 1613, and Hugh Myddelton was knighted nine years later. Myddelton died in 1631. There is a monument to his memory at Amwell Spring. By Robert Mylne, it reads: 'Sacred to the memory of Sir Hugh Mydelton, Baronet, Whose Successful Care, Assisted by the patronage of his King conveyed this Stream to London: An Immortal Work, Since men cannot more nearly imitate the Deity than in bestowing health.'

By 1703, the head of water in the Round Pond was proving insufficient to supply neighbouring properties. Accordingly, a reservoir 33ft higher was built 200yd to the north, near to where Claremont Square is now situated. Water was pumped first by a windmill and later by horse power. Daniel Defoe visited and described it in his *Tour Through England* in 1724–26, thus, 'the higher basin they fill from the lower by a great engine formerly worked by six sails, now by many horses, constantly working'. Later, in 1768, a steam engine was installed by John Smeaton. The company's expansion was rapid. In their first year they supplied 175 houses, in the fifth year 1,000, and by 1834 they were supplying 73,000 houses.

The New River Company had their first offices in Dorset Street, near Blackfriars. They moved to new premises at New River Head in 1820, transferring their Blackfriars premises to the City of London Gas Light & Coke Co. In the 1830s, the company built reservoirs at Stoke Newington and Cheshunt, the western one at Stoke Newington lined with stones from London Bridge and with a unique pumping station resembling a castle by William Chadwell Mylne. Water was no longer taken from the Round Pond in 1914, and treatment ceased altogether at New River Head in 1946.

A Multitude of Water Companies

As years passed, a multitude of water companies were founded: some lasted only a few years; some combined with larger concerns; others folded.

There were reservoirs on the high ground at Hampstead Heath, where since the 1590s the Hampstead Water Co., under the direction of the City of London, had supplied water by gravity to the St Giles area. The company later sank a well and pumped water from it to Camden Town and Kentish Town. The Hampstead Water Co. was absorbed by the New River Company in 1859.

Plans were hatched to supply spring water to Piccadilly and St James's in the early 1660s. Nothing came of the scheme, but later the York Buildings Company was formed to pump water from the Thames to the same area. Letters patent were granted in 1675 to Ralph Bucknall and Ralph Wayne, empowering them:

> … for a period of ninety-nine years to erect a water works and water house near the river of Thames upon part of the grounds of York House or York House garden, and to dig and lay ponds, pipes and cisterns for the purpose of supplying the inhabitants of St James Fields and Piccadilly with water at reasonable rents.

Water was taken from the river and pumped by a horse gin to the top of a tower. From there it was distributed by elm pipes to 2,700 houses. The York Buildings

Company was probably the first water works in the world to pump water by the power of steam – or, more accurately, by atmospheric pressure. The pioneering inventor Thomas Savery had set up shop in nearby Fleet Street, making his 'fire engines'. Savery's engine was a forerunner of those of Thomas Newcomen, inventor of the atmospheric steam engine. Unfortunately, Savery's engine was not always reliable, 'being liable to so many disorders, if a single mistake happened in the working of it, that at length it was looked upon as a useless piece of work and rejected'. It seems, however, that the problems were not entirely those of the engine, rather the solder used in its construction.

Shares in the York Buildings Company soared in the early 1700s, and at one point a £10 share rose to as much as £305. It was not to last, prompting the rhyme:

> You that are bles't with wealth by your Creator
> And want to drown your money in Thames Water,
> Buy but York Buildings, and the Cistern there
> Will sink more pence than any fool can spare.

In 1725, Thomas Newcomen himself installed an atmospheric engine at the York Buildings Company, the first in London – not that the well-to-do local inhabitants welcomed it, calling Newcomen's pioneering engine the 'York Buildings Dragon'. Eventually the coal-fired engine was taken out of service, much to the approval of the locals, who were 'very glad of it, for its workings, which was by seacoal was attended with so much smoak that it not only must pollute the air thereabouts but spoil the furniture'. Replacement atmospheric engines were later

York Buildings Company. *(By permission of Thames Water)*

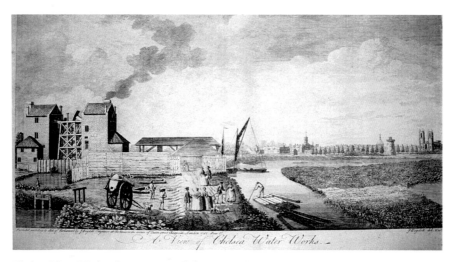

Chelsea Water Works. *(By permission of Thames Water)*

installed at the York Buildings Company in the mid-eighteenth century, to be followed by the steam engines of Boulton and Watt. Supply from York Buildings came to an end in 1825.

London was expanding eastwards and, in 1669, the speculative builder Thomas Neale (also groom porter to Charles II and builder of much of Covent Garden – hence Neal's Yard) built a water works on land he leased from the Dean of St Paul's Cathedral at Shadwell, where King Edward VII Memorial Park is now situated. It was at the Shadwell water works that Boulton and Watt, in 1778, erected their first steam engine in London for water supply. In 1807, the works were taken over by the London Dock Co.

A novel scheme, with a somewhat shady history, was pioneered by Hugh Marchant, who in 1694 used the flow of London's sewage to power a water wheel to raise Thames water. His plant was situated near Tom's Coffee House in St Martin's Lane. Marchant's water works later installed a windmill in Tottenham Court Fields.

Greenwich and the Royal Dockyards at Deptford were supplied by the Ravensbourne Water Works. In 1701, letters patent were granted 'to raise water out of Ravensbourne River in Kent and break up grounds for laying pipes for the conveniency of the inhabitants'. The works, situated on the east side of Deptford Creek, were taken over by the Kent Water Works in 1809.

The Millbank Water Works supplied the parish of St Margaret's with water taken from the Thames just upstream of the Palace of Westminster. It was taken over by the Chelsea Water Works in 1727. Supply was drawn from a tidal inlet, visible today as the Grosvenor Canal, at the point where the railway to Victoria Station crosses the Thames.

Water was raised to reservoirs in Tyburn Lane (now Park Lane) and Green Park (parallel to Piccadilly and stretching from opposite Stratton Street to Half Moon Street) where it was distributed to St James's, Whitehall and Mayfair. In 1809, the

Source of the Southwark Water Works. *(George Cruikshank)*

company began to take supplies from the Thames at Ranelagh Creek, ¾ of a mile upstream from the original works.

South of the Thames, John Strype tells us that, in 1720, 'Southwark uses chiefly the water of Thames; that falls into a great pond in St Mary Overies, that drives a mill called St Saviour's Mill. The owner is one Mr Gulston.' There were also works connected with the local Anchor Brewery at Bankend where a horse mill raised Thames water. It became known as the Borough Water Works and later, in 1820, as the Southwark Water Works.

Lambeth developed later than Southwark and its first undertaking, Lambeth Water Works, was built following an Act of Parliament in 1785. Water was extracted from the Thames at Belvedere Road and reservoirs were later constructed at Brixton Hill and Streatham Hill.

The South London Water Works was formed in 1807 to supply the newly populated areas of Newington, Camberwell and Clapham. The company had two reservoirs at Kennington Common (near to the Oval cricket ground) which were filled with water from the spring tides in Vauxhall Creek. Steam engines then raised the water to a head for distribution. The company changed its name to the Vauxhall Water Co. in 1834 and in 1845 amalgamated with the Southwark Water Co., becoming the Southwark and Vauxhall Water Co.

West London was supplied by the West Middlesex Water Works. In 1806, Robert Dodd made plans for a plant at Fulham but was overruled by William Nicholson, who built the company's first works at Hammersmith (Dodd resigned in protest). It supplied Thames water to a reservoir at Campden Hill and later extended its supply to Soho, Bloomsbury and Covent Garden. In 1838, the company built settling and storage reservoirs at Barnes. They were filled at high tide by gravity and the water was then pumped beneath the Thames in a 36in cast-iron pipe to Hammersmith.

The West Ham Water Works first took water from the River Lea in 1743, when a 'large engine worked by fire' pumped to a reservoir at Mile End. After the London Dock opened it took over both the Shadwell works and the West Ham works in 1807. Later, however, the East London Water Works built a pumping station at Old Ford and absorbed the West Ham and Shadwell undertakings from the Dock company. Initially there were four reservoirs, two on each side of the Lea, and filled at spring tides. In 1829, because industry was polluting the lower reaches of the river, supplies were taken further upstream at Lea Bridge.

The Kent Water Works, the brainchild of Robert Dodd, were founded in 1809 and supplied Deptford, Greenwich and parts of Surrey and Kent. They were quick to take over the Ravensbourne works and, on the advice of John Rennie, replaced John Smeaton's water wheel with Boulton & Watt engines. Reservoirs were eventually built at Greenwich Park, New Cross and Eltham. The company also supplied the Royal Arsenal and HM Dockyards at Woolwich and Deptford. By the mid-1850s, because supplies from the River Ravensbourne were insufficient, a bore hole was sunk at nearby Cold Bath Well and supply drawn from there. In the second half of the nineteenth century the company took over a series of smaller concerns.

The Grand Junction Water Works was formed at Brentford in 1811, with John Rennie as engineer, taking its water from the canal in the belief that it would be purer than the river. When this proved to be a mistake, the company drew water from the Thames at Chelsea and then from Kew.

Pure and Wholesome Water

At the beginning of the nineteenth century London's water was anything but pure and wholesome; the only attempt to purify it was merely to allow sediment to settle. It was left to the pamphleteer John Wright to voice Londoners' concerns, writing of the Grand Junction Company that 'several families in Westminster and its suburbs are supplied with water in a state offensive to the sight, disgusting to the imagination, and destructive to health'. He went on to say that the river at the company's intake in Chelsea was 'charged with the contents of more than 130 common sewers, the drainings from dunghills and laystalls, the refuse of hospitals, slaughter houses, colour, lead, gas and soap works, drug mills and manufactories, and with all sorts of decomposed animal and vegetable substances'. This was no exaggeration – the Gas Light & Coke Co., recently established in nearby Great Peter Street, used the Thames to deposit their effluent!

Increasing alarm induced Wright to team up with Sir Francis Burdett, the notorious radical MP (and the most popular politician of his day). They persuaded Parliament to act and a select committee was established to look into the state of London's water. It was chaired by the illustrious – but now very aged –

engineer, Thomas Telford. Recommendations were made, but they came to nothing. A rhyme in *Punch* sums up the state of the river:

King Thames was a rare old fellow,
He lay in his bed of slime,
And his face was disgustingly yellow,
Except where 'twas black with slime.
Hurrah! hurrah! for the slush and slime!

The matter from cesspools carted,
Decay'd vegetation as well;
Dogs and cats from life departed,
Sent their odours to add to the smell.
Hurrah! hurrah! for the slush and slime!

Not surprisingly, there were cholera epidemics in 1832, 1849 and 1854, but the case had still to be proved that water was a carrier of the infection. Dr John Snow made the breakthrough when he studied cases of cholera in Soho, where people were taking their water from a particular pump. In Snow's own words:

On proceeding to the spot, I found that nearly all the deaths had taken place within a short distance of the [Broad Street] pump. There were only ten deaths in houses situated decidedly nearer to another street-pump. In five of these cases the families of the deceased persons informed me that they always sent to the pump in Broad Street, as they preferred the water to that of the pumps which were nearer. In three other cases, the deceased were children who went to school near the pump in Broad Street ... With regard to the deaths occurring in the locality belonging to the pump, there were 61 instances in which I was informed that the deceased persons used to drink the pump water from Broad Street, either constantly or occasionally ...

The result of the inquiry, then, is, that there has been no particular outbreak or prevalence of cholera in this part of London except among the persons who were in the habit of drinking the water of the above-mentioned pump well.

I had an interview with the Board of Guardians of St James's parish, on the evening of the 7th inst [7 September], and represented the above circumstances to them. In consequence of what I said, the handle of the pump was removed on the following day.

Further parliamentary inquiries followed, revealing that 'water was discoloured, abounded in animal life, was unfit even for baths, was repellent for drinking purposes and was directly responsible for driving the poorer inhabitants to indulge in alcohol'.

All this activity and pressure eventually culminated in the ground-breaking Metropolis Water Act of 1852. The Act stipulated that by December 1855, all water for domestic purposes should be filtered, that all intakes of Thames water should be above the tidal limit at Teddington Lock, that reservoirs within 5 miles of St Paul's Cathedral should be covered over and that a constant supply should be provided everywhere. As a consequence, the Chelsea Water Works Co. moved their intake to Surbiton, and the Grand Junction, West Middlesex and the Southwark and Vauxhall moved their intakes to Hampton. (The Lambeth Water Works Co. had already moved their intake to Thames Ditton.)

As well as the requirement to supply unpolluted water, there were many advances in engineering in the nineteenth century. Steam pumps were now more reliable. Those of Boulton and Watt continued to be used but eventually the Cornish engine, first commissioned at the East London Water Works at Old Ford by Thomas Wicksteed, became dominant. A variety of other pumps were also introduced. In addition, there were improvements in boiler design, first with the Lancashire boiler and culminating in the water tube boiler.

In the early nineteenth century, wooden water mains gradually began to be replaced by ones of cast iron. By 1820, the New River Company had replaced their entire 400-mile network. The other companies were soon to follow, taking advantage of the properties of cast iron which allowed a greater diameter of piping, the capability of withstanding greater pressures and a lesser liability to leak.

The companies also began to give serious consideration to the purity of their supplies. At first, sediment was just allowed to settle by leaving the water to stand. Although settling made the water less turbid, it still contained bacteria and other impurities. It was James Simpson, of the Chelsea Water Works Co., who pioneered filtration to improve the quality of water. In 1829, he transferred Thames water to two settling reservoirs and then filtered it by gravity through filter beds consisting of loose bricks and tiles, above which was a layer of gravel and then, on top, a layer of sand.

The Metropolitan Water Board

Throughout the latter part of the nineteenth century, pressure was growing for London's water supply to be taken into public ownership. Impetus was given by a Royal Commission which concluded that although there was ample supply from the Thames, the 'general control of the water supply should be entrusted to a responsible public body'. There was further pressure when the London County Council (LCC) was established in 1889, and this body came close to taking control of London's water supply itself. But it was Walter Long in 1902 who introduced a bill to Parliament to establish a water board 'to manage the supply of water within London and certain adjoining districts, for transferring to the Water Board the undertaking of the Metropolitan Water Companies and for other purposes', which became the Metropolis Water Act of 1902.

Under the terms of the Act, the Metropolitan Water Board acquired the New River Company, the East London, the Southwark and Vauxhall, the West Middlesex, the Lambeth, the Chelsea and the Grand Junction companies and the Staines Reservoirs.

The Metropolitan Water Board encompassed 576 square miles from Ware in the north to Sevenoaks in the south and from Staines in the west to Gravesend in the east. By 1920, the Board had set up its headquarters at New River Head. Its sources of supply were the River Thames (69 per cent), the River Lea (15 per cent) and wells (16 per cent). Intakes from the Thames were at Laleham for the Queen Mary Reservoir, Walton and Hampton. Water was abstracted from the Lea at the New River intake between Ware and Hertford, at Enfield for the King George Reservoir and Chingford Mill for the Walthamstow Reservoir. In 1961, the Board had fifty-two well stations, twenty-eight north of the Thames and twenty-four south of it.

Water treatment continued to be based on filtration, supplemented in 1916 by the addition of chlorine (in the form of bleaching powder). Ammonia was added from 1936 to remove any unpleasant residual taste from chlorine. A system of double filtration was adopted from 1923 to counteract algal growth – water was passed first through a rapid filter bed followed by a slow sand filter.

In 1947, the Board opened the King George VI Reservoir at Staines and, in 1950, the William Girling Reservoir at Chingford.

In 1960, 51 per cent of pumping was from steam engines, 39 per cent from electric pumps and 10 per cent from oil-fired engines. The job of the pumps was (a) to pump river water to reservoirs and then to pump it out as required; (b) to pump treated water to supply lines and, if necessary, to pump again to higher ground; and (c) to pump water from wells to the surface and thence to supply.

Thames Water

By the terms of the Water Act of 1973 public water boards were reorganised. The intention was that they should cover an area corresponding to the river basin from which they took their supply, rather than being structured with political boundaries. Thus the Metropolitan Water Board was formally abolished and governance passed to the Thames Water Authority, which also took responsibility for several other supply bodies covering an area from the source of the Thames to its estuary.

The construction of the Thames Water Ring Main was a major project in the 1980s and 1990s. In 1989, the industry was privatised.

Sewage Disposal

In the early nineteenth century, the idea that impure water could be a source of disease was dismissed by most authorities. At the time, impure air was blamed – the so-called Miasma theory of ill health. But, as mentioned above, there were those, like the pamphleteer John Wright, who began to question the effects on health of impure water and, as a result, in 1828, a Royal Commission, under Thomas Telford, was set up to investigate. Despite sensible suggestions, the conclusions of the Commission came to nothing.

By the 1531 Bill of Sewers, it was illegal to discharge household effluent to sewers. Instead, foul domestic sewage, of every sort, was emptied into cesspools, dug in the ground. These were periodically emptied by 'night soil men' who carried the contents off to sell to farmers as fertilizer. Not surprisingly, cesspools leaked, overflowed and were the scene of unpleasant mishaps. In 1326, for example, poor Richard the Raker had fallen in and 'drowned in his own excrement'. The diarist Samuel Pepys had a similar experience when his neighbour's cesspool overflowed: 'Going down to my cellar, I put my feet into a great heap of turds, by which I find that Mr. Turner's house office is full and comes into my cellar.'

The situation got considerably worse with the coming of the water closet. It was Joseph Bramah, the son of a Yorkshire farmer, who improved its design. He was followed by the aptly named Thomas Crapper, who sold them in numbers, helped by his slogan 'A certain flush with every pull'. The problem was the vast increase in water flowing into the cesspools. Inevitably they overflowed, emptying their contents into the sewers and hence to the river, the major source of London's drinking water. Furthermore, since 1815, because of the inadequacies of cesspools, effluent from household drains was allowed to discharge to sewers. Thomas Cubitt neatly summed up the problem:

> Fifty years ago nearly all London had every house cleansed into a large cesspool. Now, it is carried off at once into the river. It would be a great improvement if that could be carried off independently of the town, but [instead] the Thames now is made a great cesspool.

In 1848, the Metropolitan Commission of Sewers was set up. It combined into one body the seven former Commissions of Sewers and began considering ways of improving health and drainage. Various schemes were proposed such as collecting sewage in four reservoirs and pumping it to farmland. In another, a tunnel, carrying sewage, was proposed to run from Eel Pie Island (Twickenham) to Plumstead. W.H. Smith (the newsagent) chipped in with the idea of disposal by rail. All came to nothing, but significantly, in 1849, Joseph Bazalgette joined the Commission. Then, in 1856, the Metropolitan Board of Works came into being.

Clause 135 of its constitution stated: 'The Metropolitan Board of Works shall make such sewers and works as they may think necessary for preventing all and any part of the sewage of the Metropolis from flowing into the river Thames in or near the metropolis.' John Thwaites was chairman of the newly created Board and from January 1856, Bazalgette became chief engineer. By May of the same year, Bazalgette had prepared plans for both a northern and southern drainage scheme.

Any opposition to Bazalgette's plans was conveniently nullified by the so-called 'Great Stink' of 1858. It was a hot summer and matters came to a head when Members of Parliament were prevented from carrying out their normal business by the stench emanating from the River Thames. *The Times* sums up neatly:

> The intense heat had driven our legislators from those portions of their buildings which overlook the river. A few members, bent on investigating the matter

to its very depth, ventured into the library, but they were instantaneously driven
to retreat, each man with a handkerchief to his nose.

The 'Great Stink' concentrated minds wonderfully.

Bazalgette's scheme was for a system of main sewers (which intercepted discharge from local ones) to run in a west to east direction. North of the river
there were three. The high-level one ran from Hampstead Heath to Holloway,
Stamford Hill and then to Old Ford. The intermediate sewer began in Kensal
Green, and ran via Bayswater, Oxford Street, Old Street, Bethnal Green and
thence to Old Ford. The low-level one ran from Pimlico to Westminster, the City
and Limehouse to Bromley-by-Bow. The Abbey Mills Pumping Station raised the
contents of the lower level to that of the high and intermediate levels at Old Ford
and from there all effluent was transferred to Barking where it was discharged to
the river. Problems were averted in the low-lying Fulham and Hammersmith area
by pumping from a station in Chelsea to the low-level sewer.

The pumping station at Abbey Mills is a fine building. It was designed by
Charles Driver under the direction of Joseph Bazalgette. It was originally flanked
by two tall minarets, regrettably demolished to prevent them from acting as sights
for German bombers.

The southern drainage scheme had an upper and lower level, again running from
west to east. The upper level ran from Clapham to Deptford from where it was
pumped to Crossness on the Erith marshes. The lower level ran from Putney to
Deptford with a branch coming in from Bermondsey. At Crossness, the sewage was
pumped to a covered-in reservoir and at high tide released to the river where it
would flow out to sea. Crossness was completed in 1867, and Abbey Mills one year
later. Thus for the first time the upper river was restored to a state of relative purity.

But there were still worries. Some were concerned that the pumping stations
could fall victim to an enemy attack, and the French navy were the obvious
threat. The journal *The Builder* responded reassuringly: 'The sewage of London
north and south suddenly discharged upon an advancing fleet would inevitably
produce a panic and retreat, or death by poison.'

Raw sewage was still, however, being discharged to the Thames, albeit downstream from the capital. In 1882, those who objected to this practice formed
themselves into a body known as the General Committee for the Protection of
the Lower Thames. As a result of their campaigning, a Royal Commission was set
up. It reported that 'the discharge of the sewage in its crude state … is at variance
with the original intentions and with the understanding in Parliament when the
1858 Act was passed'. Later it urged that solid material should be separated out
and sold as fertilizer, while liquid should be transferred 20 miles downstream to
Canvey Island and harmlessly discharged to sea. The cost of this proposal alarmed
the Board and instead Bazalgette suggested adding ferrous sulphate and lime to
the effluent to separate out the solids and then discharge the liquid harmlessly to
the river. Accordingly, solids were conveyed in six vessels out to sea, a process that
continued until 1998.

Abbey Mills Pumping Station, 1868. *(By permission of Newham Local Studies Library)*

Pumping station, Victoria.

5 The Thames Barrier

Under certain weather conditions London can be subject to the threat of flooding. Storm surges, arising from low pressure in the Atlantic Ocean, can cause vast quantities of water to be channelled down the North Sea and into the Thames estuary. It has happened before. In 1236, a surge tide flooded the very heart of the city and 'caused all the marshes about Woolwich to be all at sea wherein boats and other vessels were carried in the stream. In the great Palace of Westminster men did row with wherries in the midst of the Hall.' And on 7 December 1663, Samuel Pepys recorded in his diary: 'I hear there was last night the greatest tide that there was ever remembered in England to have been in this river: all Whitehall having been drowned.'

In our own times, in the East Coast flood disaster of 1953, over 300 people lost their lives. The tide rose almost 4ft above its normal level at London Bridge and was within a whisker of flowing over the parapet. Had it done so, the London Underground would have been flooded and put out of action for months.

Various schemes to overcome disastrous floods have been proposed before. In 1907, there were ill-fated plans for a barrage with locks at Gravesend. The scheme was reconsidered again in the 1930s but abandoned because of the fear of German bombing. The Port of London Authority (PLA) investigated lifting bridges, swing bridges and retractable bridges at Long Reach between Dartford and Purfleet. Professor Hermann Bondi was eventually appointed by the Greater London Council (GLC) to assess the possible mechanisms. London's Docks were thriving at the time and an obstacle to progress was the difficulty of getting large vessels through the openings of a barrier. The PLA eventually agreed to a barrier if it was situated above the entrance to the Royal Docks complex. Accordingly, a site upstream at Woolwich was chosen.

The river is about 520m wide at this point. The barrier consists of a series of rotating gates, devised by Charles Draper. It was designed by Rendel, Palmer & Tritton Ltd for the GLC. Work started in 1974. There are two types of gates. At the shallow river bank there are four falling radial gates, one on the south bank and three on the north. They do not allow the passage of river craft and drop into position when required. In the centre there are six rising sector gates, four of 61m length and two of 31.5m length. When open, they sit flush with the river and allow river vessels to pass through. When notice is given of a surge tide, the sector gates are rotated into position and together with the radial gates (which are dropped into position) a 20m-high steel wall is created to prevent water flowing upstream. To protect land downstream, suitably high river walls have been built.

The Thames Barrier is operated by the Environment Agency. It was put into commission in 1982 and protects 125 square kilometres of London from flooding. It can be closed in one and a half hours in response to hydrological and meteorological data received at its control room.

Part 2
MANUFACTURING

6 Bell Founding

As long ago as 1420, Robert Chamberlain was making bells in Aldgate, so beginning a long tradition in this area of London. Then, in 1570, Robert Mot began an uninterrupted tradition of bell making at the Whitechapel Bell Foundry that continues to the present day.

Robert Mot inscribed his bells with the words 'Robert Mot made me', or its Latin translation '*Robertus Mot me fecit*'. It was not long before Mot was making bells for Westminster Abbey. Two bells were made in 1583 and 1598, both inscribed '*Campanis Patrem laudate sonantibus altum*' ('Praise the High Father with sonorous bells'). Thus began an association with the abbey that is still maintained today.

In 1605, Mot sold the business to Joseph Carter and then, in 1617, there began a long period when the bell foundry was owned by the Bartlet family. It was at this time that other bell foundries sprang up in London, but as an inscription on a bell at Richmond in Surrey testifies, it would seem that those from Whitechapel were superior: 'Lambert made me weak not fit to ring, But Bartlet amongst the rest hath made me sing.'

Richard Phelps and Thomas Lester took over the Whitechapel Bell Foundry in 1700, and made over 350 bells for churches in the City of London. This included the 5¼-ton clock bell for St Paul's Cathedral, dated 1716, and which still gives good service to this day.

Throughout the nineteenth century, the Mears family ran the foundry at Whitechapel and it was during this time that its most famous bell was cast – Big Ben, the foundry's largest bell at 13½ tons. Other famous bells were Great Peter for York Minster and Great Tom for Lincoln Cathedral, as well as bells sold overseas, including America's Liberty Bell. In 1919, George V and Queen Mary visited the foundry to witness the casting of new bells for Westminster Abbey.

The technique of bell founding has changed little throughout the ages. Bells are made from an alloy of copper and tin in the ratio of 4:1 and cast into moulds. The mould that forms the inside of the bell is known as the core, and that which forms the outside the cope. They are both made from London clay, and when perfectly smooth and suitably inscribed both are clamped together. Molten metal from a furnace is then poured into the space between the two. The mould is finally broken apart after cooling to reveal the finished product.

Bell casting at Whitechapel Bell Foundry. *(By permission of Tower Hamlets Local History Library)*

The skill of the founder is nicely summed up by the inscription on the bell at Ingatestone in Essex:

The founder he has played his part,
Which shows him master of his art,
So hang me well and ring me true,
And I will sound your praises due.

The Whitechapel Bell Foundry is at 32–34 Whitechapel Road, London E1 1DY (www.whitechapelbellfoundry.co.uk). Tours of the foundry are held by arrangement.

7 Candle Making

Candles have been a source of light for well over 2,000 years. The discovery of ancient candlesticks – there is part of an Etruscan bronze candlestick in the British Museum dating from the fifth century BC – provides evidence of their antiquity. In the Middle Ages candles for general lighting purposes were made of tallow, obtained from the fat of beef or mutton. For the church, and for lawyers' seals, the more expensive beeswax was used. Monks would make the candle by pouring hot beeswax down a long wick suspended from the ceiling, allowing it to adhere, and then repeating the process.

Trade in London was regulated by one of the two livery companies: the Wax Chandlers and the Tallow Chandlers. The Wax Chandlers are twentieth in order of precedence of the City livery companies, date from 1330 and received their first charter in 1484. Their hall is in Gresham Street. The Tallow Chandlers received their first charter in 1462 and are twenty-first in order of precedence. Their hall is in Dowgate Hill, part of which dates from a post-Great Fire rebuilding of 1672 and still survives today.

Much to the dismay of the Tallow Chandlers, oil lamps began to become popular in the seventeenth century. They objected with much vigour to the Lord Mayor and Aldermen that they:

... thought themselves safe from any other invasion of their laws, one of which is an Act of Council, ... whereby every household from the first of October to the first of March in every year for ever should cause a substantial lantern and a candle of eight in the pound to be hanged without their door, ... notwithstanding the said act ... certain unfree men have set up ... a great number of lamps and other lucidaries which are merely novel. It is humbly hoped that projects so pernicious to the public good of the kingdom ... will not find or receive any encouragement in this city.

In the eighteenth and nineteenth centuries other products were discovered for making candles. Spermaceti, a waxy substance extracted from whale oil, made good candles. Also, by saponification – a chemical process used in the manufacture of soap (whereby a metallic alkali reacts with a fat or oil) – stearine was produced, which made excellent candles that burnt with a brighter light and with less smoke and smell than tallow.

Price's Candles

William Wilson was born in Scotland. He came to London following the failure of his father's ironworks, working as a merchant with his partner, Benjamin Lancaster, importing goods, including tallow, from Russia. Then, in 1830, they founded E. Price & Co. and began to make candles. (Price was the surname of Lancaster's aunt but it seems she had no involvement with the business.) They were among many small firms making candles in London, but by the end of the century Price's became the largest candle manufacturer in the world.

Wilson and Lancaster were quick to acquire the rights to exploit a patent for the separation of the solid and liquid constituents of coconut oil and began making candles from a mixture of refined tallow and coconut oil. They opened a factory at Vauxhall (the Belmont Works) and another in Battersea, and purchased a coconut plantation in Sri Lanka. William Wilson's son, George, was an excellent chemist. He experimented with the technique of steam distillation and was thus able to make candles from many raw materials, such as skin fat, bone fat, fish oil and industrial greases. Candles were also made from palm oil – extracted from palm nuts harvested in Ghana, Nigeria and Togo. Palm nuts became the basis of the company's seal, which depicts Africans bringing the oil to a figure of Britannia beneath a palm tree.

The company expanded rapidly. In 1849, it bought the night light business of Samuel Childs and soon opened a factory on Merseyside at Bromborough to take advantage of the fact that all imports of palm oil came into Liverpool Docks. The Wilson family were evangelical Christians and the Bromborough plant became a model of good practice. Houses were built for the workers and the village so created was later copied by Lever's at Port Sunlight and Cadbury's at Bournville. Meanwhile, the factory at Vauxhall was disposed of and production in London concentrated at Battersea, which then took the name Belmont.

In the meantime, George Wilson was continuing with his research and was successful in isolating glycerine as a by-product of palm oil. This invaluable viscous and hygroscopic liquid found (and still finds) many valuable uses in pharmaceuticals and other industries. For his efforts George Wilson was awarded a Fellowship of the Royal Society (FRS). Wilson also isolated oleine, the first lubricating oil marketed by Price's. They were later to develop the enormously successful Motorine, the Rolls-Royce of lubricants, so termed because it was used in all Rolls-Royce cars.

Price's Candles, Battersea.

The development of the petrochemical industry was to radically change many industries, including candle manufacture. Fractional distillation of crude oil enabled paraffin wax to be isolated, which was to replace stearine as a raw material. But candles were soon to become a rather specialised form of lighting – gas light and the new electric light predominated by the end of the nineteenth century. Nevertheless, the market for church candles remained and Price's acquired the candle manufacturer Francis Tucker, founded in South Kensington and later moving to Putney, which supplied the Catholic Church. They also bought Charles Farris, located first in the City and then in Hounslow, which made candles for Anglican churches.

Lever Brothers took over Price's in 1919, and in 1922 created, with Shell, BP and Burmah Oil, Candles Ltd. Lubricating oils and candles continued to be made at Battersea and in 1951 the company produced the first multigrade oil: Energol. As part owners of Candles Ltd, BP transferred production to their Grangemouth refinery and so brought to an end the development and manufacture of lubricating oils at Battersea. BP and Burmah in 1982, and Shell in 1991, sold their shares in Candles Ltd and Price's once again became a private company, concentrating on its core and traditional business of candle making.

By 2001, Price's had left Battersea and are now based in Bicester and Bedford, but the name is still synonymous with candle manufacture in the UK. Captain Scott took a supply of Price's stearine candles on his ill-fated expedition to the South Pole. Remarkably enough, the candles could, in cases of emergency, be eaten. Such was the case when a group of coal miners managed to survive for fourteen days by eating tallow candles after a mining disaster trapped them underground. Always associated with the royal family, Price's hold four royal warrants and supply their candles for all state occasions, including the funeral of Princess Diana.

8 The Chemical Industry

Xylonite

In 1846, the German-Swiss chemist Christian Friedrich Schönbein accidentally spilled some concentrated nitric acid on his kitchen table. He reached for a cloth to wipe it up, which he then placed near the fire to dry. The cloth turned out to be made of cotton and dry it did; but it also exploded with a violent flash. Schönbein had inadvertently formed nitrocellulose (also known as cellulose nitrate), a highly flammable and explosive material. When dissolved in ether or alcohol it was known as collodion and was used as a photographic film base, applied as a liquid film onto a glass support.

But it was left to the English metallurgist Alexander Parkes to exploit the discovery of cellulose nitrate and invent what proved to be the first thermoplastic material in the world. He combined cellulose nitrate with camphor and found a solid material remained after evaporating off the solvent. He patented his new material and exhibited it at the 1862 International Exhibition in London. He found he could make many useful household items – buttons, combs, knife handles – from his new material, which he called Parkesine.

Parkes worked in Birmingham for the firm of Elkington's but it was Hackney in London that saw the birth of the plastics industry. He founded the Parkesine Company and opened a factory at Wallis Road in Hackney Wick in 1866, next door to the works of George Spill & Co., which manufactured rubber-coated waterproof capes and groundsheets for the British Army. Spill's brother, Daniel, joined Parkes and became works manager.

However, within a couple of years the firm failed. Parkesine shrank and distorted, probably because of insufficient camphor. In addition, the price had been set too low to make it profitable. Parkes then left the scene and went back to Birmingham to continue his career as a metallurgist. But Daniel Spill persevered. He changed the name of the product to xylonite (xylon was taken from the Greek word for wood) and the company name to the Xylonite Co. Ltd. He later took out a lease at 122 & 124 Homerton High Street and called his premises the Ivoride Works (xylonite proved to be an excellent substitute for ivory) and made knife handles, brooches and insulating material for electric cables.

But trouble was brewing for Daniel Spill in the USA. There, John Wesley Hyatt was making a similar product, which he called celluloid. One of its uses in America was as a coating for billiard balls. Much to the alarm of Wild West saloon keepers, every time two balls struck each other there was a loud explosive bang and all cowboys in the joint pulled their guns. More seriously, Spill sued Hyatt for infringing patents rights. At first he was successful but eventually judgement went against him, the verdict being that the patent did not belong to him but to Parkes. Daniel Spill subsequently joined forces with Charles Pearce Merriman, who was making combs in the building adjacent to his premises, and a contract was later agreed with the Impermeable Collar & Cuff Co. of Bower Road, Hackney Wick, to make collars and cuffs laminated with xylonite; in other words, 'wipe clean collars and cuffs'.

Fire was an ever-present hazard, however, and in 1883, the Metropolitan Board of Works declared the factory unsafe, forcing the directors to stump up £2,500 for improvements. Two years later, a serious fire forced a move out of Homerton and in 1887, a new factory for xylonite production was built at Brantham in north Essex; the unmoulded product was shipped back to Homerton where the final products were made.

The company had many well-known customers, such as Boots, which was supplied with combs, hair pins, tooth and shaving brushes and toys. In the early twentieth century the Homerton factory closed and production was moved to Hale End in Walthamstow.

Paint and Varnish

Prussian blue – the first stable pigment to find common use – was first prepared by Heinrich Diesbach and Johann Konrad Dippel in Berlin in 1704. It was sometimes known as Berlin blue and takes its name from the vivid blue uniforms of the Prussian Army.

Berger's Paint

Louis Amelius Christianus Adolphus Steigenberger, a colour chemist living in Frankfurt-on-Main, did not discover Prussian blue but he perfected an easy way to synthesise it. He was 19 at the time and in 1760 moved to London to set up as a manufacturer of colours. He went into partnership with a Mr Rapp and so jealous were the pair of their secret formulations that the deed of their partnership included a clause stating, 'each partner should at any time provide a private room in either of their dwelling houses for secret consultations'. Anglicising his name to Lewis Berger, he married a local girl, Elizabeth Alger, set up home first in Ratcliffe, then Shadwell, and built a factory to manufacture colours. By 1790, Berger was selling nineteen different colours and had opened a new factory in Morning Lane, Homerton, where the family eventually moved. The firm's trademark was the god Mercury, a reference to one of their colours, vermilion, the toxic chemical mercuric sulphide. There was a range of other colours as well, such as carmine, manufactured from the cochineal insect, and chrome. The latter was made from chrome ore which was crushed before formulation by a horse gin, causing a heavy roller to revolve over it.

The House of Berger supplied artists' colours. Many were bought by Mr J. Newman of Soho Square, whose shop was patronised by many famous artists. Berger can lay claim that his colours were used in many 'old masters' – none more so than 'Lucerne Lake' and 'The Fighting Temeraire' by J.M.W. Turner. He also sold colours to John Doulton, head of the famous Lambeth pottery, for the world-renowned Royal Doulton.

Berger had a shop and distribution centre in Well Court, off Cheapside. Colours made in Homerton were dispatched there by wagons from whence they were distributed all over the country by coach. The firm provided colours to Messrs Cooper, sealing wax makers, who had premises on London's medieval bridge. When the bridge was demolished in the 1820s they moved to St Bride Street to trade in vermilion. Interestingly, John Berger (Lewis' son) purchased the cobble stones from the old bridge's roadway and laid them at the Homerton works between the clock tower and the workmen's entrance. It is unknown where they are now.

In 1850, the North London Railway Co. built a viaduct through the firm's land and was forced to pay £1,000 in compensation. The railway arches found a use as stables and for storage. Lewis Curwood Berger and Capel Berger succeeded their father, John, in the 1860s. Lewis had previously run a business in Southall making copper sulphate, an ingredient, with arsenic, of the colour emerald green. And it was at this time that Berger's diversified to make paints. Well Court eventually closed and all operations from then on were carried out at Homerton. In the early twentieth century Berger's employed over 400 staff. Many employees had long periods of service, but there was the occasional strike, one in 1911 causing the company to be accused of having 'sheds of death'. In 1934, Berger's purchased a site in Freshwater Road, Chadwell Heath, and began making nitrocellulose-based paints, Homerton eventually closing in 1957. After many mergers and takeovers, Berger's became Crown Paints.

Thomas Parsons & Sons

London's eighteenth-century coach builders and coach ironmongers were located in and around Covent Garden; George Shillibeer, pioneer in 1829 of London's first omnibus service from Paddington to Bank, worked there. Nearby, at 5 Long Acre, were the paint and varnish manufacturers Edward Wood and William Innell. Wood and Innell supplied their varnish exclusively to the coach trade and in the early nineteenth century took on George Parsons, younger son of one of their best customers. Parsons was born in Bath in 1790 and his father, John, was in the coaching business: 'John Parsons' flying wagons set out from his warehouse in Walcot Street, Bath, every Wednesday and Saturday night; arriving at the New White Horse Cellar and Black Bear, Piccadilly, and White Hart, Basinghall Street, and return to Bath and Bristol in the same time.' But it was not the coaching business that George entered – it was as a varnish manufacturer with Edward Wood.

Varnish has been known since ancient times. It has the useful property of drying to form a hard lacquer which is insoluble in water and durable. The Egyptians prepared it by dissolving resins in oil. They coated their wooden mummy cases with it and there are many examples on display at the British Museum. Varnish was also known to the Chinese, who tapped the liquid sap of the varnish tree (Tsi-chou) to obtain a milky emulsion. It was also used in Asian japanning art work.

Manufacturers of varnish guarded their secret formulations very jealously; when Parsons mixed his ingredients he forbade anyone to look over his shoulder to see what was going on. He was also in the habit of spitting into the brew to test its consistency. In 1811, the firm opened a small factory in Battle Bridge (now King's Cross), described at the time as 'bone stores, chemical works and potteries, render[ing] it peculiarly unsavoury'.

Another factory was opened in Mitcham in the mid-nineteenth century. At the same time, Parsons moved from 5 to 40 Long Acre, but when Covent Garden Underground Station was built he was forced to move to 8 Endell Street nearby. The firm also had premises at 313–315 Oxford Street. After the Second World War the headquarters moved to Mitcham.

A.F. Suter & Co.
The firm was founded by Albert Frederick Suter in the early twentieth century, manufacturing shellac, a resin extracted from certain trees from India and Thailand. It had many uses including as a sealant and for the manufacture of 78rpm gramophone records. The firm was at Eastway in Hackney Wick and then Dace Road, Bow.

William Perkin and Mauveine
In common with so many scientists, William Perkin was inspired by lectures given by the great Michael Faraday. He recorded that as a young boy of 14 in 1852, he 'sat up in that gallery an eager listener to lectures by that great man [and] I little thought then that in four years' time I should be the fortunate discoverer of the mauve dye'. And so it was that in a back room in Cable Street, in London's East End, Perkin synthesised the world's first organic dye, at the tender age of 18.

Perkin, the youngest son of a carpenter, was educated at the City of London School, then in Milk Street, and, according to his father, was set for a career as an architect. Instead he fell under the spell of chemistry, writing, 'a young friend was good enough to show me some chemical experiments; amongst these were some on crystallisation, which seemed to me a most marvellous phenomenon: as a result, my choice was fixed and it became my desire to be a chemist'.

In 1853, Perkin went to work for the great August Wilhelm Hofmann at the Royal College of Chemistry (later the Royal College of Science at Imperial College) in Oxford Street. Hofmann had given him the task of synthesising quinine, a remedy for malaria, and so enthusiastic was the young Perkin that in the evenings and in any spare time he could find, he worked away in a crude laboratory he had set up in a back room of his home in Cable Street. There he oxidised aniline (containing toluidine impurities) with potassium dichromate and got a messy black solid which, to his surprise, when dissolved in alcohol gave a purple solution capable of dyeing fabric and remaining colour-fast even when exposed to bright light.

Perkin was quick to realise the importance of his discovery. Up until then, natural materials were used as dyes but they lacked brilliance and definition of colour. Perkin first got in touch with a dyeing firm in Scotland but then – with money put up by his father – set up himself in a factory at Greenford on the banks of the Grand Union Canal. Perkin's raw material was coal tar, available in abundance as a by-product from London's gasworks. His discovery heralded a new industry. *Mr Punch* explains:

> There's hardly a thing that a man can name
> Of use or beauty in life's small game
> But you can extract in alembic or jar,
> From the physical basis of black coal tar:
> Oil and ointment, wax and wine,
> And the lovely colour called aniline:
> You can make anything from a salve to a star
> (If you only know how) from black coal tar.

Perkin patented his discovery in 1856, and more synthetic dyes were to follow: alizarin, aniline pink, magenta and others. The Greenford factory thrived – it was said the colour of the Grand Union Canal was often mauve – but Perkin always retained an interest in fundamental, laboratory-based research and in order to concentrate on research he sold his commercial interests to Brooke, Simpson & Spiller in 1874. They were based at Atlas Works in Berkshire Road, Hackney Wick. It was here that Arthur George Green discovered the dyestuff primuline. The eminent British chemist Raphael Meldola worked for a while in Hackney Wick and discovered the dye Meldola's blue.

William Henry Perkin – the founder of the coal tar industry – was knighted in 1906.

Carless, Capel & Leonard

In 1859, Eugene Carless opened the Hope Chemical Works, an oil-refining business in White Post Lane, Hackney Wick. Carless' distillery imported crude oil from America and refined coal tar and shales to make benzoline, paraffin oil and carburine. In 1872, the partnership of Carless, Capel & Leonard came into being and in 1895 bought the nearby Pharos Chemical Works. By 1893, the firm was marketing petroleum, used as a fuel for motor launches. At the suggestion of Frederick Simms, an early motoring pioneer and associate of Gottlieb Daimler, they called their product 'petrol'. Although not accepted as a trade name, the word 'petrol' passed into common use. The firm supplied their fuel for the London to Brighton run of 1896, and both Simms and Leonard were founder members of the AA and RAC. At the beginning of the twentieth century the firm had a virtual monopoly of the production of petrol. They purchased the Lea Chemical Works in 1907. New refineries were opened outside London, and production ceased in Hackney Wick in the 1970s.

Chemical Manufacture

Ferrous sulphate heptahydrate was known as copperas (confusingly, as it contains no copper). It was used in ink manufacture and as a dye. In 1678, Daniel Colwell described his manufacturing process in the Royal Society's journal *Philosophical Transactions*. He had premises at Deptford and produced copperas by oxidation and hydrolysis of iron pyrites. To achieve this he dug a 12ft-deep trench (100ft by 15ft) and lined it with wood. Iron pyrites were added to a depth of 2ft and allowed to oxidise and hydrolyse by the action of the atmosphere and rain. The process was lengthy; it was called 'ripening' and took up to five years. To give a measure of continuous production, fresh pyrites were added each year until the trench was full. Copperas liquor was conveyed from the bottom of the bed by a wooden trough which led to the boiling house. Here the liquor was boiled and crystals of copperas formed on twigs. Sulphuric acid was a by-product and had to be removed; to test for its presence, the ingenious device of popping a boiled egg into the liquor was employed. If the eggshell dissolved, bits of scrap iron were added to remove excess sulphuric acid. Copperas was used to dye top hats, an Act of Parliament forbidding them to be dyed in any other way.

Sulphuric acid was employed in a vast array of industries: by bleachers, refiners, japanners, in the alkali and dyestuffs industry, in the manufacture of plastics and pigments, etc. Its manufacture started in Britain at the works of the notorious quack and swindler Joshua Ward of Twickenham. He was perhaps better known for his pills and potions, efficacious, if at all, because of their opium content, and used by such notables as George II, Horace Walpole and Henry Fielding. Others were more sceptical:

Before you take his drop or pill,
Take leave of friends and make your will.

He opened in Twickenham in 1736, aided by John White, 'the ingenious chymist who carried on the Great Vitriol Works at Twickenham'. Rows of 50-gallon round-bottomed flasks, with a small quantity of water in each, were secured in sand, and sulphur and potassium nitrate added. The mixture was ignited by plunging a red-hot iron into the flasks. After the reaction had subsided, fresh materials were added. It is recorded that the 'fumes from the chimneys and the general stench aroused much irritation among the neighbourhood', causing Ward to move to Richmond. As an alternative to Ward's flasks, lead chambers were pioneered by John Roebuck in Birmingham in 1746, and in 1772 were taken up by the firm of Kingscote & Walker in Battersea – the so-called lead chamber process.

It has been claimed – rather unexpectedly – that a country's wealth can be gauged by the amount of sulphuric acid it produces. There were seven vitriol plants in London by 1820, surprisingly more than in any other place in the country.

The contact process for the manufacture of sulphuric acid came to London in 1870. It relied on the catalytic conversion of sulphur dioxide to sulphur trioxide and subsequent dissolution to form sulphuric acid. In 1876, Rudolph Messel and W.S. Squire oxidised sulphur dioxide over platinised pumice in Silvertown. Sulphuric acid was also prepared at the Crown Sulphur Works in Marshgate Lane, Stratford, at the West Ham Chemical Works in Canning Road, and nearby by Thomas Bell & Co.

Agricultural fertilizer went by the very appropriate name of chemical manure. It was made by reacting calcium phosphate with sulphuric acid to give the slightly soluble calcium superphosphate, which was then mixed with sawdust or other dry absorbent. Bones were the first raw material as a source of phosphates. They were used as fertilizers in the early part of the nineteenth century but it was soon realised that they worked only on acid soils, the acid releasing superphosphate.

John Bennet Lawes was the founder of the chemical fertilizer industry. He was born in 1814 and inherited the family estate at Rothamsted in Hertfordshire. (Rothamsted Research is now the largest agricultural research centre in the UK and almost certainly the oldest agricultural research station in the world.) He was newly married to Caroline Fountaine. He was also looking for a site for his factory. Caroline had been promised a European honeymoon; instead, Lawes hired a boat and took the poor girl down the Thames to Deptford Creek. Here, he fixed on a site for his factory. Perhaps Caroline took comfort from the fact that her husband became very rich on the proceeds of chemical manure. Lawes also opened the Atlas Chemical Works in Millwall and another factory in Barking. In the mid-nineteenth century, the manager at J.B. Lawes' factory was Joseph Kemball; in 1870, he left to set up his own factory at Bromley-by-Bow to make citric acid and tartaric acid. Citric acid was produced from calcium citrate imported from Sicily, and tartaric acid from the deposits on the inside of wine barrels. In 1931, the Italian government – no doubt to gain a monopoly on its production – banned the export of calcium citrate. Kemball, Bishop & Co. (as they were then known) made a deal with the American firm Pfizer to import calcium citrate from them. Production continued at Bromley-by-Bow until 1971.

Silvertown was an early centre of the chemical industry. As mentioned in Chapter 3, the area takes its name from Stephen Winckworth Silver who opened his India Rubber, Gutta Percha & Telegraph Works Co. there in 1852. Soon to follow was Odam's Chemical Manure Works, which used the blood of freshly slaughtered cattle from its own abattoir – the stench must have been dreadful. In 1896, the Mineral Oils Corporation (Minoco) was formed when Charles Hunting began to distil and refine lubricants from Russian crude oil imported by the parent company, the Northern Petroleum Tank Steamship Co. of Newcastle upon Tyne. Five years later, the company was known as Silvertown Lubricants and from a 13-acre site supplied oil to railway companies. In 1929, it was taken over by Gulf Oil. By the 1960s, distilling had given way to oil blending.

Next door was the chemical works of Brunner Mond. They were at Crescent Wharf in Silvertown and had two plants, one making soda crystals from ammonia and the other caustic soda. In the Great War they were asked to convert their plant to purify TNT as part of the war effort. Brunner Mond were unwilling

Silvertown Rubber Works, 1936. *(By permission of Newham Local Studies Library)*

Brunner Mond Silvertown explosion. *(By permission of Newham Local Studies Library)*

at first, as well they might have been, for within 400yd of the works were 3,000 residents, a school and a church. However, they eventually agreed to the government's demands and production started. The process consisted of dissolving 5-ton batches of TNT in a melt pot with alcohol, then crystallising and packing into 50lb cotton bags. On Friday 19 January 1917, a fire started in the melt pot room and the largest ever explosion experienced in London was soon to follow. The appalling Silvertown explosion killed 73 people and 2,000 were made homeless.

The blast, as 50 tons of TNT exploded, was heard as far away as the south coast. The cause of the explosion was never discovered – some said it was sabotage, but poor safety procedures were more likely. The explosion was a disaster waiting to happen. The company's chief scientist, F.A. Freeth, wrote that 'it was manifestly very dangerous. At the end of every month we used to write to Silvertown to say that their plant would go up sooner or later, but were told it was worth the risk.'

Howard & Sons

Luke Howard was born in Clerkenwell to Robert Howard, a tinsmith, and it was Robert who encouraged his son to take up a career in chemistry. At the time Luke was apprenticed to a pharmaceutical chemist in Stockport. His father wrote: 'Chemistry is a noble science and becomes useful in many sorts of business as well as a lasting source of amusement.' On Luke's return to London, Robert Howard set him up as a retail chemist at 29 Fleet Street, opposite the church of St Dunstan in the West. Luke Howard was fortunate to marry into money and this enabled him in 1797 to enter into partnership with William Allen of Plough Court and Plaistow. Allen and Howard synthesised chemicals. The partnership lasted for ten years, after which William Allen continued at Plough Court (eventually to become Allen & Hanburys), while Luke remained at Plaistow (Howards Road) before moving to Stratford and taking on Joseph Jewell, his foreman, as a partner. They employed nine men in 1813. The firm made a range of chemicals including borax, copper carbonate, arsenic sulphide and, importantly, ether.

Joseph Jewell was something of a poet as well as a chemist. I dare say he was a better chemist than a poet, as the following illustrates:

And then the aether which I made,
Claimed nearly all the aether trade
Throughout the trading nation
And I with confidence might say;
That none before me knew the way.

Robert Liston, the renowned Scottish surgeon, pioneered the use of ether as an anaesthetic in 1846 at University College Hospital. Whether or not he obtained it from Howards is unknown, but either way they found themselves in an excellent position to exploit this new market. A letter sent to them in 1847 from Edinburgh explains all: 'Gentlemen: We request that you will send per first steam vessel if possible a bottle of sulphuric ether, … it is … to be inhaled by patients requiring operations … the effect being to prevent them … from feeling pain.'

Howards began making quinine in 1823, and this was to become their major product throughout the nineteenth century. They also made cocaine, caffeine and morphia, but soon gave up this side of their business after the visit of a Home Office official who was horrified to discover a specimen of opium 'the size of a polo ball'. When the business was threatened with prosecution, the offending substance was thrown into the furnace.

The firm acquired Hopkin & Williams in 1889, and in 1897 bought a site at Ilford. They completed the move to Ilford in 1923. Howards were eventually taken over by Laporte.

May & Baker

John May, born in 1809 in Harwich, was apprenticed to an Ipswich chemist and druggist before moving to London to work for Charles John Price in Battersea, whose firm was described as 'the chief manufacturing house of sulphate of magnesia, mercury preparations etc. in the south of England'. There were other chemical plants in Battersea manufacturing sulphuric acid and dyestuffs, and before long, John May set up in business with Joseph Pickett and Thomas Shipp Grimwade at a site to the west of Battersea Bridge. Pickett died within the year and Grimwade, 'not caring for cooking chemicals', retired. This prompted May to take on William Baker, who had served an apprenticeship under his father at Chelmsford, as a partner in 1840. So was born the firm of May & Baker. In 1841, they bought Garden Wharf at Battersea and remained there until 1934. May & Baker exhibited at the Great Exhibition in 1851 and won many prizes for their pharmaceuticals, 'all of which appeared to be of excellent quality'. John May and William Baker both lived in Battersea, May at Hyde House in Hyde Terrace, and Baker at The Cedars in Battersea High Street.

In time, May & Baker obtained additional premises at Bell Lane in Wandsworth. In 1922, they were taken over by the French company Rhône-Poulenc, and in the 1930s moved from Battersea and Wandsworth to Dagenham in Essex. The Battersea site was sold to the Morgan Crucible Co., originally set up by the six Morgan brothers as the Patent Plumbago Crucible Co., making graphite crucibles. They changed their name to the Morgan Crucible Co. in 1881, by which time they were the largest crucible manufacturers in the world. In 1904, they diversified and began making carbon brushes, soon expanding their operations overseas. The Battersea Works was the film set for one of the Oscar-winning Ealing comedies, *The Man in the White Suit*, starring Alec Guinness. In 1979, the works began to wind down and within five years all manufacturing was moved to South Wales and Worcestershire.

By the late 1930s, May & Baker had discovered the sulphonamide antibacterial sulphopyridine. And it was as well that they did, for in 1942 it was successful in treating Winston Churchill's pneumonia. He was grateful and said: 'This admirable M&B, from which I didn't suffer any inconvenience, was used at the earliest moment and, after a week's fever, the intruders were repulsed.' And it was not only Churchill. In 1944, Nero, a circus lion, was similarly cured. The Dagenham site eventually combined with Hoechst in 1999 and became known as Aventis.

Allen & Hanburys

The well-known pharmaceutical company of Allen & Hanburys can trace its roots to a Welsh apothecary and Quaker, Silvanus Bevan. He and his brother Timothy began at Old Plough Court in Lombard Street in 1715. William Allen

Allen & Hanburys, Bethnal Green, in the late nineteenth century. *(By permission of Tower Hamlets Local History Library)*

Weighing and packing goods at Allen & Hanburys in the 1930s. *(By permission of Tower Hamlets Local History Library)*

joined the firm in 1792. Allen was connected through marriage to the Hanbury family who took control on Allen's death.

By the mid-nineteenth century, they had factories at Ware in Hertfordshire and Bethnal Green in East London. It was at Bethnal Green that all research was carried out and pharmaceutical preparations and surgical instruments were made. Allen & Hanburys pioneered the drug salbutamol in 1968, a mainstay of asthma treatment. The company was absorbed by Glaxo in 1958, and the Bethnal Green premises closed in 1982.

Johnson Matthey

The story of the multinational company Johnson Matthey begins with Percival Norton Johnson. He was born in the City of London in 1792, son of John Johnson, an assayer of ores and metals. Percival was apprenticed to his father and in 1814 gained the freedom of the Worshipful Company of Goldsmiths. But he was clearly interested in chemistry and wrote a paper entitled 'Experiments which prove platina, when combined with gold and silver, to be soluble in nitric acid'. By 1817, he had taken over the business and in 1822 signed a lease at 79 Hatton Garden.

Johnson knew how to prepare platinum in a malleable form and wished to refine metals and fabricate them into small articles. In this he was helped by Thomas Cock who used to work for the firm of William Allen in Plough Court (see above). To begin with, Johnson had only one employee, James Hamlet; and so began a succession of Hamlets who together completed 120 years of continuous service.

In the 1820s, Johnson began working with nickel silver. The preparation of the finished product – 18 per cent nickel, 55 per cent copper and 27 per cent zinc – involved Hamlet going to a 'remote spot' in Bow where Johnson had erected a shed with a reverberating furnace. Here, out of harm's way, he sintered the raw material to drive off the arsenic. (Johnson was evidently ahead of his time, for in 1844 the Metropolitan Buildings Act came on to the statute book. It sought to protect populated areas – in practice the well-to-do West End of London – from the new noxious industries of the nineteenth century. These were almost always chemical works, and it was preferred that they should be sited in the marshy area to the east of London. In the words of Charles Dickens, the area was 'a place of refuge for trade establishments turned out of town'.) Nickel silver was sold by Johnson to the manufacturers of knives, forks and spoons.

George Stokes joined the business in 1826 and later became a partner, as did George Matthey in 1851. Johnson Matthey became official assayers and refiners to the Bank of England. The firm expanded and leases were taken at 78 Hatton Garden and 23 Leather Lane. Mellish's Sufferance Wharf was also acquired in Millwall on the Isle of Dogs where a steam engine and machinery were installed to crush and grind gold quartz, silver, lead and other ores. By the 1950s, manufacturing operations were being gradually withdrawn from Hatton Garden.

Starch Manufacture

Starch, an important ingredient of the human diet and with many industrial applications (such as paper making), consists of glucose molecules bonded together. It is extracted from the seeds and roots of plants. J. & J. Colman (the well-known mustard manufacturer based in Norwich) had a factory in Bethnal Green. It began as the Bethnal Green Starch Works and was taken over by Colman's in the 1860s. It was situated in Old Ford Road near its junction with Cambridge Heath Road.

Soap

Soaps are salts of fatty acids, made by treating vegetable oils or fats with an alkaline solution (often sodium hydroxide). Saponification occurs when the fats are hydrolysed to soap.

Edward Cook was a soap manufacturer in Norwich. In the 1830s, his firm moved to Goodman's Yard in Whitechapel and then, in 1859, to Cook's Road in Bow on the banks of the River Lea in a factory known as the 'Soapery'. Edward Cook's son (also Edward) was a distinguished chemist and for a time MP for West Ham. Many women worked in the soap-packing department, the most popular brand being 'Throne Toilet Soap', said to have a fragrance similar to a Devonshire cottage garden on a lovely May morning. After the Second World War, Cook's were taken over by Knight's. Knight's Castile soap was introduced in 1919 by John Knight Ltd. Initially at Wapping, they moved to Silvertown at the Royal Primrose Soap Works and are now part of Unilever. The original 'Castile' soap was made in Castile, Spain, from olive oil and soda. In 1919, the firm employed 1,200 people.

Hawes Soap Works were sited just to the west of Blackfriars Bridge on the south bank of the Thames. The firm occupied a building called the Old Barge House and in the nineteenth century were the largest soap makers in London. Benjamin Hawes was a government minister and was married to Marc Brunel's daughter.

John Jeyes patented a disinfectant fluid in 1877, and in 1885 set up the Jeyes' Sanitary Compounds Co. in Richmond Road, Plaistow, to manufacture it. In 1903, Yardley had a factory in Carpenters Road, Stratford, which by 1937 extended to the High Street. The firm moved to Basildon in 1966, but their distinctive logo 'The Flower Sellers' remains on their building in Stratford High Street.

Andrew Pears, son of a Cornish farmer, moved to London in 1787 and worked as a barber in Gerrard Street, Soho. As a sideline he made cosmetics and then a gentle soap for his upper-class clients. Pears opened in Oxford Street in 1789, and won prizes at the Great Exhibition of 1851. Their motto was 'Any fool can make soap, but it takes a wise man to sell it'. John Everett Millais' painting 'Bubbles' was – with his permission and with a bar of soap added – used by Pears in one of their advertisements. The firm's name was also included on the back of the penny lilac stamp in 1881 and the halfpenny vermilion in 1887. Pears eventually

Knight's Royal Primrose Soap Works, Silvertown. *(By permission of Newham Local Studies Library)*

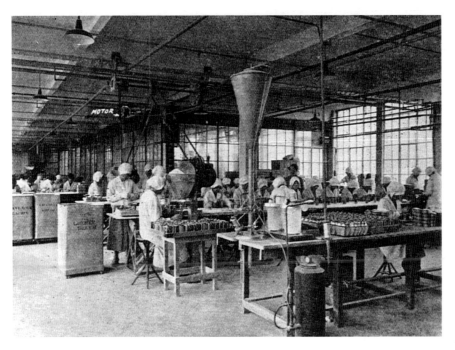

Yardley's soap factory, Stratford. *(By permission of Newham Local Studies Library)*

moved to Isleworth, near to the railway station, in 1862, and were later taken over by Unilever.

The Chiswick Soap Co. had works in Burlington Lane, opposite Chiswick Square, in 1878. They made Cherry Blossom boot polish and moved to a large site in the triangle between Burlington Lane, Great Chertsey Road and Hogarth Lane. They were eventually taken over by Reckitt & Colman, and in 1972 production moved to Hull.

9 Clockmaking

Early clocks tended to be made by blacksmiths. Thomas Tompion, the 'father of English watch making', was the son of a Bedfordshire blacksmith. In 1671, he arrived in London, became a member of the Clockmakers' Company, and set himself up in business in Water Lane, off Fleet Street. Tompion soon made the acquaintance of Robert Hooke, Curator of Experiments for the Royal Society, and the mathematician Sir Jonas Moore, Surveyor-General of Ordnance at the Tower of London. No doubt because of his friendship with Moore, Tompion was commissioned in 1674 to make a turret clock for the Tower of London. He also made watches for Charles II and William III, and clocks for John Flamsteed at the newly founded Royal Observatory.

Tompion was succeeded by George Graham who was first apprenticed to Henry Aske, a lantern clockmaker. Graham had married Tompion's niece and went into partnership with him in 1711. Tompion died in 1713, and a few years later the business, at the sign of the 'Dial and Three Crowns', moved 100yd down Fleet Street. Meanwhile, John Harrison was busy perfecting watches to enable longitude to be calculated at sea. He lived for a while in Orange Street, off Red Lion Square, and devoted years of his life to inventing a means of telling the time at sea in order to calculate longitude. His problems centred on the disturbance the movement of the waves would have on conventional timekeepers and also the changes in temperature. His efforts were eventually successful but the Board of Longitude, which had offered a prize of £20,000 for a solution to the problem, was not immediately convinced. Graham was a great admirer of Harrison, giving him much support and encouragement in his quest for recognition. Eventually, George III intervened and Harrison got his due reward.

There were many Huguenot clockmakers in London, including Nicholas Urseau. He was living in Westminster in 1568 and made the magnificent clock in Clock Court at Hampton Court. He was succeeded by Bartholomew Newsom, also clockmaker to the Queen. The large number of foreign clockmakers in London induced sixteen English clockmakers to petition the King to grant them recognition, resulting in a charter being granted to the Clockmakers' Company in 1631.

Edward East had a workshop on the corner of Ram Alley and Fleet Street. He was watchmaker to both Charles I and Charles II. It is said that as he walked across St James's Park to his execution, Charles I gave 'his precious silver alarm clock of Mr East' to Sir Thomas Herbert.

Thomas Earnshaw was born in Ashton-under-Lyne in 1749. He worked for John Brockbank and Thomas Wright, watchmakers in Poultry, and in 1781 was successful in improving the design of chronometers. There was a delay in patenting Earnshaw's work, by which time John Arnold, another well-known watchmaker, had claimed and registered the design as his own. Embittered by this, Earnshaw accused Brockbank of divulging his secret to Arnold. In 1798, Earnshaw set up at 119 High Holborn. He was also awarded money by the Board of Longitude.

John Vulliamy was born in Switzerland in 1712 and came to England to join Benjamin Gray, clockmaker to George II, at his Pall Mall workshop. He married Gray's daughter and inherited his business. The office of clockmaker to the king remained in the Vulliamy family until the death of Benjamin Lewis Vulliamy in 1854, at which point the firm came to an end.

Clerkenwell was a centre of the watch- and clock-making industry in London. T.E. White has described the area as it was in the early part of the twentieth century in an article entitled 'A Clerkenwell Tour of Fifty Years Ago', published in *Antiquarian Horology*. John Smith built a factory in St John's Square in 1834. It was the largest manufactory in Clerkenwell and exhibited at the Great Exhibition in 1851. There were many subdivisions of the trade: escapement maker, engine turner, fusee cutter, springer, etc. A Mr Pullen was a fine dial writer and his workshop was filled with dials of every sort. The Beesley family were in Clerkenwell Road and their business was founded there in 1796. Also in Clerkenwell Road was a Mr Cox, who did watch repairs, and the premises of Robert Pringle & Sons at 'Wilderness Works', the largest horological trade house in Britain. Nearby was the Synchronome company of Frank Hope-Jones. The firm made Shortt-Synchronome pendulum clocks supplied to observatories all over the world. Known as Father Time, Hope-Jones pioneered the pips time signal broadcast by the BBC.

Switzerland has always been a centre of the watchmaking industry. Louis Victor Baume and Pierre Celestin Baume were Swiss watchmakers from Les Bois. In 1844, Celestin arrived in the capital to found a London house of the firm, beginning in Clerkenwell and moving to Hatton Garden in 1887. John Moore & Sons was founded in about 1790 and had a factory in Clerkenwell Close. They made all sorts of clocks for 100 years, as well as wind dials, weather vanes, roof ventilators and lighthouse equipment. They had a workforce of eighty in 1835. Thomas Mercer was born in St Helens in 1822. He moved to London and began making marine chronometers. The firm was situated in Goswell Road, Clerkenwell. There were many other watch and clockmakers in the area. The 1861 census lists scores, turning out over 100,000 watches every year.

10 Engineering

Famous London Engineers

Joseph Bramah

Joseph Bramah was born in 1749 in Stainborough near Barnsley. His early life was spent as a farm labourer, probably on land owned by Thomas Wentworth, 3rd Earl of Stafford, for whom his father worked as a manservant. An accident to his ankle put paid to his farming job and Bramah began an apprenticeship with Thomas Allott, a local carpenter and joiner. After he had completed his apprenticeship, he made his way to London on foot and set up as a cabinet maker in Cross Court, Carnaby Market, near Golden Square in Soho.

Bramah is best known for his innovative work on the water closet and the hydraulic pump. The principle of the water closet was discovered by Sir John Harrington in 1596 and patented by Alexander Cummings, a watchmaker working in Bond Street. Although similar to the water closet of today, 'it was said to stink' and never found much application. It was Bramah, with his technical mind, who improved the closet. His innovation was to connect the handle that releases the contents of the pan to the wire that opens the valve and releases the flush of water from the cistern above. Matthew Boulton was certainly impressed and promptly ordered one at a cost of £11 2s 3d for his new house in Birmingham. The water closet proved to be a great success and by the end of the nineteenth century, 6,000 of them had been produced, Bramah meanwhile moving his workshop to Denmark Street, St Giles.

In response to a spate of house burglaries, Bramah invented a new type of lock, which was an improvement on the tumbler lock. In his own words, the 'idea of constructing a lock was suggested to me by the alarming increase in house robberies'. He published a paper on its construction and such was the demand that he soon realised he should mechanise its construction. This is where Henry Maudslay enters the story. He went to work for Bramah at the age of 18 and it is very likely that he made a major contribution to Bramah's locks. (In 1798, Bramah offered 200 guineas to anyone who could open his locks without the key. Alfred Hobbs, a locksmith, spent fifty-one hours trying and eventually succeeded.)

Bramah's agile mind enabled him to invent many other devices, including the hydraulic press, steam engines and pumps, the slide rest, a numbering machine for bank notes, and a 'new method for making pens'. In all, he took out eighteen patents. One was for a hydraulic machine for uprooting trees. He was trying it out in Holt Forest, Hampshire, when he caught a cold. It turned into pneumonia which proved fatal. Bramah was buried in Paddington churchyard in 1814.

Henry Maudslay

Perhaps the most notable of London's early engineering concerns was the one founded by Henry Maudslay. He was born in Woolwich in 1771 and began life as a powder monkey, later in a carpenter's workshop and then at a forge. His mechanical dexterity came to the notice of Joseph Bramah, well known for his water closet and hydraulic press. But Maudslay fell out with Bramah over money. He asked for a rise and was turned down. The falling-out prompted Maudslay to set up on his own, first in Wells Street, off Oxford Street, and then in Margaret Street, Marylebone.

In 1800, Marc Isambard Brunel commissioned Maudslay to construct machines to make ships' pulley blocks, used for rigging sails of naval vessels. There was an enormous demand for pulley blocks, which at that time were made individually. Portsmouth Dockyard ordered forty-five of Maudslay's machines and a production line was set up to produce the blocks. Before Maudslay's invention, over 100 men were employed making blocks by hand. Now a team of ten could turn out 160,000 every year. Maudslay also worked with Bryan Donkin of Bermondsey on a mechanism for driving lathes and in 1807 patented his Table Engine, a steam engine suitable for driving factory machinery.

Maudslay left Margaret Street in 1810 and moved to a derelict riding school in Lambeth. There the firm was to remain at premises in Westminster Bridge Road. He went into partnership with Joshua Field, his draughtsman, and the firm became, in due time, Maudslay, Sons & Field Ltd. Maudslay built many marine engines. His first was in 1815 for the Thames riverboat *Richmond*. In 1823, he constructed the engine for the Royal Navy's first steam-powered vessel, *Lightning*.

His greatest contributions to engineering in general were his slide rest, screw-cutting lathe, and machine tools — machines to make machines. The slide rest made it possible for him to construct the block-making machinery for Brunel, to produce multiple identical pieces of work. Before his invention of the screw-cutting lathe, every screw had to be produced by hand and have its own individual nut — a very lengthy business. The lathe brought mechanisation, greater speed of manufacture and improved accuracy.

Maudslay also made a micrometer so that measurements could be made with the highest degree of accuracy. It was in constant use in his workshops, so much so that his workmen referred to it as 'the Lord Chancellor' — an obvious reference to its good judgement. The firm's output was prolific. Coining machines were made for the Royal Mint and the mints of overseas governments, and stationary steam engines were provided for the London & Blackwall Railway. Marc Brunel's tunnelling shield for his Rotherhithe Tunnel was Maudslay's work, as was the time ball mechanism for the Royal Observatory at Greenwich. But it could be argued that Maudslay's greatest offering was the men he trained — Joseph Clement, James Nasmyth, Joseph Whitworth and many others — all of whom went on to make their own contributions to Britain's pre-eminence in nineteenth-century engineering.

Joseph Clement

Joseph Clement was the son of a Westmorland handloom weaver. He arrived in London in 1813 and went to work for Joseph Bramah. When Bramah died, his sons took over the business and they soon fell out with Clement who then joined Henry Maudslay as his chief draughtsman. He later started on his own in Prospect Place, Newington Butts.

Clement is well known for his role in the manufacture of Charles Babbage's ill-fated Difference Engine. Babbage was fascinated by numbers and determined to invent a calculating machine. At the time, complicated calculations were printed in books of tables which, unfortunately, often contained errors. Such errors had disastrous consequences: for example, in 1707, four naval ships under the command of Sir Cloudesley Shovell were wrecked off the Isles of Scilly because he had calculated longitude with error-strewn tables. Babbage employed Clement – a brilliant engineer – but the pair fell out because of Clement's high costs. The Difference Engine was never completed and Babbage died a disappointed man. (In another context, there is the story of the whistle Clement made for Isambard Kingdom Brunel's Great Western Railway. It was six times over budget and Brunel complained. 'My whistle is six times better,' retorted Clement. 'If you order a first rate article you must be content to pay for it.')

Bryan Donkin

Bryan Donkin was born in Northumberland, the third son of a well-to-do surveyor and land agent. For a while he worked for the Duke of Dorset at Knole Park, Sevenoaks, and then, following the advice of John Smeaton, was apprenticed to the eminent engineer John Hall. In 1798, he set up on his own, making moulds for hand-made paper. The concern was financed by Hall, whose sister-in-law Donkin married.

It was the Frenchman Didot Saint-Leger who, in 1799, took out a patent in France for a paper-making machine. The idea came to the attention of a wealthy firm of stationers in London, the Fourdrinier brothers, and they set up a factory in Bermondsey with Donkin in charge to produce the machines. After prolonged trials at the Two Waters Mill in Hertfordshire, Donkin succeeded in his endeavours and the Bermondsey firm went on to produce 191 machines, 83 of which were sold in Britain. There were other inventions: taps, dies and bar lathes, all made in Bermondsey at the firm which Donkin now owned. In 1808, Donkin turned his attention to replacing the quill pen with one made of steel. To achieve this he welded two thin sheets of metal together, selling the pen for 3s 6d and the nibs for 1s.

Then, in 1812, Donkin perfected a process for 'preserving animal and vegetable food for a long time from perishing or becoming useless'. He did this by pioneering the humble tin can and so aided Britain's armed forces (and others) to preserve food and not have to salt, smoke or desiccate it. Despite the fact that the only way of opening Donkin's cans was to hit them with hammer and chisel, his invention proved an immediate success. Overseas explorers took canned products with them on their voyages, such as the tinned beef and pea soup taken to the Arctic on HMS *Fury* in 1824. Cans were found 113 years later and their contents were still edible. Sir John Franklin used them on his ill-fated voyage in 1845 to navigate the North West Passage. They were found years later, still as good as ever.

Later, Donkin invented a printing machine in which the type blocks or plates were fixed to a polygon which rotated each face into position for an impression to be made – the first step towards a rotary machine. He also developed a machine – originally patented by Sir William Congreve (well known for his work with solid fuel rockets) – for printing stamps and bank notes. And, so that the exact number of stamps or notes could be recorded, he invented a rev-counter. The Bermondsey works also made water wheels and sold them to Thomas Telford, who was described by Donkin as 'a very good friend for giving us work'.

By the mid-nineteenth century, the Bermondsey factory was one of the finest in the country. It was continued by his sons, John, Bryan and Thomas, after Donkin's death in 1855. The firm then concentrated on supplying the gas industry, its first contract being rack and pinion valves for the Gas Light & Coke Co. Towards the end of the century, the firm experimented with powdered coal, a precursor to pulverised fuel, before moving to Chesterfield in 1902.

John Rennie and Albion Mills

In the late nineteenth century the Birmingham firm of Boulton & Watt took on the then unknown John Rennie to install their first rotative steam engines, recently patented by Watt, to power millstones at Samuel Wyatt's Albion Flour Mills. Twenty pairs of millstones were able to grind ten bushels of wheat every hour of the day and night. This was far greater productivity than was possible with traditional water and wind power. Many millers lost their livelihood and Albion Mills were the talk of London. They even became a tourist attraction and were visited by Thomas Jefferson, who was at that time American ambassador to France. Open for just three years, they were destroyed by an enormous fire on the night of 3 March 1791. Some say the fire was caused by arson and indeed millers were soon to be seen dancing for joy on Blackfriars Bridge. *The Star* reported that 'millers within 30 or 40 miles of London have been in a state of intoxication since Wednesday last'. Rennie and Wyatt, however, believed the fire had been started accidentally by the friction of badly lubricated machinery. A popular song of the day sums up:

> And now the folks begin to chat,
> How the owners did this and that.
> But very few did sorrow show,
> That the Albion Mills were burnt so low.

Rennie was just one of a number of celebrated engineers associated with Southwark. In 1791, he set up a business of his own in nearby Stamford Street. His first interest was canals and he was involved with fen drainage. He is best known, however, for dock and bridge building. Apart from his docks in London he worked at Hull and Liverpool. He designed the first Southwark Bridge as well as the first Waterloo Bridge – described by the famous Venetian sculptor Antonio Canova as the 'noblest bridge in the world, worth a visit from the remotest corners of the earth'. Praise indeed! The business was continued by his sons and it was John (later Sir John Rennie) who completed his father's London Bridge in 1831. As well as building stationary and marine engines, the firm also constructed a number of railway locomotives. In the 1850s, they opened a shipyard at Deptford Creek.

David Kirkaldy

David Kirkaldy was another Southwark engineer with an international reputation. Like Rennie he was a Scot, born in Dundee in 1820. His early working life was at Robert Napier's Vulcan Foundry in Glasgow, but in 1861 he set out on his own to design a machine to test construction materials placed under various types of stress – namely pulling, thrusting, bending, twisting, shearing, punching and bulging. Greenwood & Batley of Leeds built his testing machine and this was soon installed at The Grove in Southwark, next door to the engineering firm of Easton & Amos.

Southwark Street was soon to be cut through and it was to here, No 99, that Kirkaldy moved in 1872. His firm had an international reputation but at a more local level Joseph Cubitt engaged him to test the materials of construction for the new Blackfriars Bridge. The works were sold in 1965, but the original machine is still in place at the Kirkaldy Testing Museum set up in 1983. It can apply a load of 300 tons and was used to test materials from the Tay Bridge after its failure in 1879.

Kirkaldy Testing Museum, Southwark Street.

Engineering Works

James Easton came to The Grove early in the nineteenth century and soon went into partnership with Charles Amos. In 1844, they built the waterworks for the newly opened Trafalgar Square. They were well-known manufacturers of beam engines and exhibited at the 1851 Great Exhibition. One of their beam engines is preserved at the Kew Bridge Steam Museum, having worked for some eighty years at a pumping station in Northampton. Another is preserved at the Westonzoyland Pumping Station Museum near Bridgwater, after sterling work over the years draining the Somerset Levels.

The City lead works of Grey & Marten were just east of Southwark Bridge, where the *Financial Times* is now situated. The works were demolished in 1980. The firm imported lead, mainly from Australia. Before the Second World War it was unloaded from lighters at the riverside wharf. In the 1920s, lead sheet was produced by melting the imported pig in pots underneath the arches of the bridge and then allowing it to run out onto a bed of 7-ton capacity to give lead sheet up to 9in thick. Lead pipes were made by pouring the molten metal into a cylinder. After cooling, a ram was forced down the cylinder by a vertical hydraulic press to extrude the lead in pipe form, which was then wound onto a drum. Lead is a highly toxic metal and, after the war, demand fell for its use in pipes and sheet, the firm then diversifying to manufacture lead solder.

There was a shot tower and lead works at 63 Belvedere Road, Lambeth. It was erected in 1826 for Thomas Maltby & Co. and was taken over by Walkers, Parker & Co. in 1839. The shot tower east of Waterloo Bridge remained until 1948 and then made way for the Festival of Britain site.

In Southwark was Edward Hayward, senior partner in the Union Street firm of Hayward Brothers; the firm had moved from their first premises at 'The Sign of the Dog's Head in the Pot' in Blackfriars Road. In 1871, Hayward invented and patented a revolutionary system for illuminating basements. The patent, entitled 'Improvement in Pavement Lighting', employed a clever use of optics, not only to allow light to pass through the glass-covered pavement, but also to direct it in an inclined direction to the basement below. This was done with a series of glass prisms beneath the pavement, so arranged as to direct the incident light in the required direction. As well as pavement lights the firm made a number of other products, such as coal hole covers, iron staircases and collapsible steel gates for lifts.

Engineering companies abounded in Southwark. Willcox's were in Southwark Street making a variety of products, including patented wire-bound hose. They specialised in supplying the agricultural industry and exhibited at many agricultural shows. Moser's, opposite the church of St George the Martyr, Borough High Street, was started by Thomas Bunnet at the sign of the 'Old Anchor and Key'. Joined by Richard Moser in 1814 and financed by the wealthy South Wales ironmaster William Crawshay, their trade at the time was in tools, locks, horseshoes, nails and household utensils. Their attitude to staff was typical of the late Victorian era. In the 1870s, the yardmen were given an extra 2*d* per week on pay day to buy beer and tobacco so that they need not break into their wages. Everyone earning more than 30s a week was paid no more, the surplus being kept by the firm to be given to the men as and when needed.

James Adams was another inventive man based in Southwark. One day in 1882 he was lying in bed with a painful attack of sciatica and feeling even worse because his bedroom door had not been shut properly, resulting in a continual rattling. This got on his nerves so much that he determined to do something about it. Thus was born his pneumatic door spring hinge or slave, patented in 1890 as 'Improvements in and in connection with door closing apparatus'. The firm of James Adams & Son started in Union Street but later moved to 26 Blackfriars Road.

The nineteenth century saw an enormous expansion of industry in south London, reflecting the vast population increase and its need to be supplied with goods. In 1890, the booklet *South London Illustrated* was published and in it are detailed scores of small concerns operating in the Bankside area. There were manufacturers of dyes, hot water radiators, disinfectants, chemicals, scales and weighing machines, boots and shoes, sacks and tarpaulins, photographic dark rooms, black lead, dustbins and surgical instruments to name just a few, as well as umpteen foundries and general engineering works.

David Napier – Engine Manufacturer

David Napier was born in 1785 in Inveraray, a son of the Duke of Argyll's black-smith. At the age of 23 he made the sea journey to London (it was quicker that way) and started work for Henry Maudslay. Napier, cousin of the Millwall ship-builder also called David Napier, set up his own business at Lloyd's Court, St Giles, and began making printing machinery. One of his customers was T.C. Hansard, the parliamentary reporter, who purchased a 'perfecting machine' from Napier, capable of printing on both sides of the paper in a single operation. Hansard was well pleased, describing it as 'this beautiful mechanism' and 'infallible' – it produced 2,000 impressions every hour, 1,000 sheets printed on both sides. Napier sold many printing machines, some steam-driven, others operated by hand. By now he had moved to larger premises at York Place, Lambeth, near to where the Festival Hall stands today, and in the mid-nineteenth century was employing between 200 and 300 men.

Napier made many machines. In the late 1830s, one of his machines was capable of making 25,000 bullets per day for the Board of Ordnance. The Board also purchased cranes, steam engines and lathes. A coin weighing machine was made for the Bank of England, and Isambard Kingdom Brunel bought hydraulic presses for his Great Western Railway at Bristol to lift wagons from line to line.

David Napier died in 1873, by which time the business was being run by his son, James Murdoch Napier. The firm specialised in machinery for government departments, the Bank of England, the Royal Mint and overseas mints. But James Murdoch Napier was a difficult man. He fell out with one of his best customers, Charles Fremantle, Deputy Master of the Royal Mint. Fremantle had ambitions for the Mint and wanted to 'pull the place together'. As part of this exercise he went with James Napier on a ten-week tour of foreign mints, but it was not a harmonious trip and Fremantle added fuel to the fire by offering one of Napier's workmen a job at the Mint. The upshot was that the furious Napier refused to speak to Fremantle, resulting, not surprisingly, in a cessation of orders. Things went from bad to worse, and by the end of the nineteenth century the firm was in serious decline.

In 1895, the 25-year-old Montague Stanley Napier took over the business and realised at once the need to diversify. He began by making machine tools for the cycle industry. But then came the motor car and the intervention of an extrovert Australian, Selwyn Francis Edge. Edge worked for the Dunlop Tyre Co. and following a visit to Paris wrote of his 'Motoring Reminiscences' in that city. He vowed to 'advance the cause of automobilism and awaken all in this country to the possibilities of this enormous new industry' which was, at that moment, in its cradle. He bought a Panhard-Levassor car (there is one on display in the Making of the Modern World gallery at the Science Museum) and determined to improve it. Edge's car had solid tyres, a tiller steering and an inefficient engine. But there was no real improvement. Edge, a salesman, needed an engineer to fulfil his dream and this is where Napier came in. He designed a new engine, replaced the tyres with pneumatic ones and the tiller with a steering wheel.

Napier's first engine had two cylinders, one firing after the other and both connected to the crank shaft. So there were two firing strokes followed by two strokes when neither cylinder fired. *The Autocar* magazine was impressed: 'It is refreshing to be able to add yet another name to the list of the English motor builders, whose work, when closely scrutinised, so excels the products of the French usines.' Edge, now in business at his Motor Vehicle Co. in Regent Street, bought six cars from Napier before 1900, increasing to 396 by 1904. The cars sold for about £500. It was an arrangement copied later by C.S. Rolls (the salesman) and Henry Royce (the engineer).

To publicise the car it was entered, along with sixty-four others, in the 1,000-mile trial organised by the Automobile Association in 1900. In April all competitors left Hyde Park Corner to travel the length and breadth of the country – up the west coast via Carlisle to Edinburgh and down the east coast via Newcastle. Despite its clutch slipping so the car could not be stopped from running in the reverse direction down hills, the reverse gear jamming and only being freed 'with the persuasion of a crow bar' and the carburettor being fixed with copper wire and plaster of Paris, the Napier car came first in its class and second overall – success indeed. Once again *The Autocar* was full of praise, writing, 'in no sense whatsoever, did it suggest a horse'.

A later car (four-cylinder, 80bhp at 800rpm) set an unofficial speed record of 66.93mph over 5 miles, with a speed of 86.25mph within a flying kilometre in France. The man from *The Autocar* had never seen the like. He wrote:

> ... words altogether fail; I cannot pretend to a verbal description of the 'big demon' as she hurtled by, more like a projectile from a big gun than a self

propelled vehicle steered and controlled by a human hand. She grew from the appearance of a little truck on four wheels as she came over the top of the hill to a shrieking monster as she whistled by. The air forced from her path hit me hard in the face and chest, stone, sticks and leaves fled after her in a vortex, and a great cloud of dust rose up instantly and cloaked her from view.

More races were entered, including those organised by the American newspaper owner James Gordon Bennett, whose purpose was to challenge French domination. In 1902 the Napier triumphed.

Expansion forced a move from Lambeth to Stanley Gardens in Acton in 1902, and by 1906 the firm was employing 1,000 men and making about 200 cars every year. It was at Acton that Edge designed a six-cylinder engine, a major innovation in the motor industry. But Napier never embraced mass production and so could not compete with Morris who made 5,000 cars every year. *Bicycling News and Motor Review* commented: 'the man of moderate means is not the class of client for whom the Napier motor is intended. It is essentially a rich man's luxury.' And distinguished clients it had: the Viceroy of India, Lord Northcliffe and Winston Churchill. Napier's did attempt to make a cheaper model, 'for doctors and others who require a car which must be absolutely relied upon under all conditions, but to whom high speed and great power are secondary considerations'. It cost £270. But the First World War brought disaster to Napier's luxury model. The company became a 'controlled establishment' under the Defence of the Realm Act and the Munitions of War Act, and instead of cars made vans, ambulances and aero engines. And it was an elderly Montague Napier who, in 1918, made the decision to cease making motor cars and concentrate instead on aero engines. 'I do not think it is so much a question of our giving up the motor trade as it will be of the motor trade giving us up.'

The Bentley Motor Car

It was in Conduit Street, in well-to-do Mayfair, that W.O. and H.M. Bentley, together with Henry Varley (recruited from Vauxhall) and F.T. Burgess (from Humber), conceived the design for the first Bentley motor car. Not far away in New Street Mews (now Chagford Street) just off Baker Street, Bentley had a garage where he sold the French DFP 2-litre motor car and it was here that the first 3-litre Bentley engine was built in 1919. At the time, most motor cars were manufactured in Birmingham, near to the suppliers of their metal parts, but Bentley wanted his cars made in London so he could keep an eye on what was going on. When a site in Cricklewood came onto the market he quickly bought it and built a factory. It was situated on the Edgware Road, just south of the Welsh Harp reservoir where the North Circular Road is today.

Bentley's finances were far from healthy and the firm was fortunate that Woolf 'Babe' Barnato, one of the so-called 'Bentley Boys', a wealthy social set, took a keen interest and went on to effectively buy the company in 1925. He had raced Bentleys at Brooklands and obviously felt a strong affection for the make. But then came the Wall Street Crash – Bentley left Cricklewood and the firm was bought by Rolls-Royce.

The Vauxhall Motor Car

The story began in Vauxhall when, in 1857, Alexander Wilson, a Scottish engineer, began making marine engines at his Vauxhall Ironworks in Wandsworth Road. He employed 150 workers who fitted engines to vessels including the pleasure boats *Queen Elizabeth* and *Cardinal Wolsey* plying between Westminster and Hampton Court. Wilson left Vauxhall to set up in Fenchurch Street as a consultant engineer, leaving W. Gardner in charge of what was now a limited liability company. He was joined by F.W. Hodges, an enthusiast for the newly pioneered motor car. Hodges purchased a car, took it to pieces and by so doing learnt how to make one. The first Vauxhall car was put on the market for £150 in 1903; it had no reverse gear! A reverse gear was added one year later, and a three-cylinder model could be bought for £375. Vauxhall sold seventy-six and then moved to Luton. The rest is history.

AEC

The famous Routemaster buses were made by the Associated Equipment Co. (better known as AEC). The company grew out of the London General Omnibus Co. (LGOC) which for a short while in the early twentieth century made buses for its routes in London at a factory in Blackhorse Lane, Walthamstow. In the First World War AEC made lorries for the war effort. Production was moved to Southall in 1927.

JAP Engines

John Alfred Prestwich was born in 1874 in Kensington. Engineering was his passion and by the age of 14 he had made a miniature steam engine. After schooling at the City & Guilds School and the City of London School, where he excelled at mathematics and as a draughtsman, he went to work for Ferranti in Charterhouse Square, making electrical apparatus. Then, at the age of 20, he set up on his own making electrical equipment in a greenhouse in his father's back garden in Tottenham High Road. At this time, motion pictures were in their infancy and Prestwich began making cinematographic equipment, cameras and projectors, etc., for a brief time in partnership with the cinematograph pioneer William Friese-Greene.

In 1903, Prestwich turned to making motor cycle engines and a 293cc JAP engine was introduced to the market. It was an immediate success and was quickly taken up by the Triumph Cycle Co. But Prestwich's interests were not confined to motor cycles. He wanted to design engines for the newly invented aeroplane and so travelled to Le Mans to witness the Wright Brothers' flying machine in action. Back in Tottenham and in collaboration with a local flying pioneer, H.J. Harding, he manufactured a monoplane which was flown over Tottenham Marshes, much to the displeasure of local farmers who complained because the noise of the monoplane scared their horses. More trouble was to follow when Prestwich entered into a brief partnership with A.V. Roe. They rented premises beneath railway arches on Clapton Marshes and flew at four o'clock in the morning. The police put an end to it, accusing the pair of 'causing a disturbance'.

By 1911, the motor cycle industry was expanding rapidly and this persuaded Prestwich to concentrate on making engines for them. At the same time he moved to a new factory at Northumberland Park, Tottenham, and was then engaged in war work. Next door an associate company was set up to make pencils, appropriately called Pencils Ltd. By the middle of the twentieth century it was making 1½ million pencils every week.

Prestwich sold his engines for use in tractors, ploughs and the manufacturing industry. A branch factory opened in 1942 to avoid concentrating all production in one place and so reduce the likelihood of bombing putting the firm out of business. After the war, Prestwich's engines powered the Cooper JAP racing car – often seen (and heard) at Silverstone and Prescott Hill climbs. In 1945, the enterprise was taken over by Villiers. Northumberland Park closed in 1963.

Charles A. Vandervell

Charles A. Vandervell moved from Willesden to Worple Way, Acton, in 1904. The firm made electrical equipment and in the early twentieth century pioneered a means to illuminate double-decker buses with lighting produced from dynamo-charged batteries. Known as CAV, they were taken over by Joseph Lucas in 1926. By 1980, 3,000 were employed at Worple Way making electrical equipment for commercial vehicles. In 1927, Charles A. Vandervell purchased the O. & S. Oilless Bearing Co. of Victoria Road. He put his son, G.A. (Tony) Vandervell, in charge and the new firm was known as Vandervell Products, making thin-wall engine bearings in a factory on Western Avenue, designed by Sir Aston Webb. In the 1950s, the company made bearings for racing cars. They also made their own racing car, the Vanwall Special. Vanwalls were spectacularly successful, winning many Grand Prix races. Vandervell's was eventually taken over by GKN, and the Acton works closed in 1970.

Aircraft Manufacture

People have always yearned to fly. Before the principles of flight were understood, men, logically enough, copied birds. Early flapping devices were known as ornithopters and many broken bones resulted when their intrepid pilots jumped from hills or buildings. An alternative approach was pioneered by the Parisian Montgolfier brothers, who realised that if a balloon is filled with air lighter than its surrounding air, it will float or rise. In 1783, they demonstrated this principle by flying a duck and a cockerel over the rooftops of Paris. The occasion was witnessed by the French King, who made the observation that criminals could be sent up! But it was Sir George Cayley who first applied scientific principles to flight. Known as the 'Father of Aviation', he defined the forces acting on an aeroplane as lift (depending on the wing shape); thrust (from the engine); weight (of the aircraft); and drag (air resistance). Cayley's work led to the development of gliders and before long the first powered flight was made, by the Wright brothers on 17 December 1903, at Kitty Hawk, North Carolina.

Handley Page

Frederick Handley Page was an early English pioneer of flight. He was born in Cheltenham in 1885 and qualified as an electrical engineer. In 1906, he was appointed chief engineer at Johnson & Phillips, a firm of electrical engineers at Charlton in south-east London. But his real passion was aviation and while still at Johnson & Phillips he began experimenting with aircraft. It got him the sack. Handley Page was not easily put off and soon set up on his own at Woolwich in a shed. But for his prototype biplanes and monoplanes he needed a flying ground and this he found north of the river between Barking Creek and Dagenham Dock. In 1909, he founded Britain's first public company that manufactured aeroplanes – it is said he raised the money by winning at poker. Facilities at Barking proved insufficient and in 1912 Handley Page moved to Cricklewood and tested his aircraft at an adjacent airfield.

In the First World War, London came under attack from German Zeppelin raids, prompting the Admiralty to ask Handley Page to make a 'bloody paralyser of an aeroplane'. This was the twin-engine O/100 bomber. Later, a four-engine bomber, the V/1500, was produced with the intention of bombing Berlin. However, the war ended before it was put into action. After the war, Handley Page turned to passenger transport to carry travellers the short distance from London to Paris. This arm of Handley Page's operations later merged with others to form Britain's first national airline, Imperial Airways.

But air travel was still far from a safe mode of transport. To improve safety, the Handley Page slat was introduced, a slot running along the wing. It was quickly taken up by other companies. In 1929, a new airfield was acquired at Radlett. Handley Page made aircraft in the Second World War: the Hampden and the famous Halifax, a four-engine bomber. Frederick Handley Page was knighted in 1942, but the company resisted mergers recommended by the government and went into liquidation in 1970.

De Havilland

Geoffrey de Havilland, son of a clergyman, was born in Hampshire and educated at Rugby School and St Edward's School, Oxford. His first interest was motor cycles which he built at the Crystal Palace School of Engineering. Like so many others, he caught the aviation bug and with a loan from his grandfather designed an engine for an aeroplane he had bought from the Iris Motor Co., of Scrubs Lane, Willesden. Renting a workshop in Fulham, he constructed his own biplane, but when tested on the Hampshire Downs it crashed. Not to be put off, he made another biplane which this time successfully flew for ¼ mile.

De Havilland became the chief designer at George Holt Thomas' Aircraft Manufacturing Co. (Airco) at Hendon, but when Airco sold up to Birmingham Small Arms Co. (BSA), de Havilland decided to set up on his own. Thus, in 1920 was formed the De Havilland Aircraft Company. He rented a field at Stag Lane, Edgware, and built civil aircraft. The company later moved to Hatfield. Their famous aircraft included the Tiger Moth, Gypsy Moth and the De Havilland Comet, the first commercial jet airliner.

Short Brothers

Eustace and Oswald Short, later to be joined by Horace Short, rented some railway arches in Battersea in the early twentieth century and set up as Short Brothers. They were contracted by the Wright Brothers to build six Wright Flyers and so became the first aircraft manufacturing company in the world. In 1909, they moved to the Isle of Sheppey. In 1913, Shorts were joined by Charles Richard Fairey as chief engineer, but he left in 1915 to set up the Fairey Aviation Company on his own.

Fairey Aviation Company

Charles Richard Fairey was born in Hendon, went to the Merchant Taylors' School and studied electrical engineering and chemistry. He worked for a while as an analytical chemist at a power station, but his love of model aeroplanes and their construction led to his appointment as chief engineer at Short Brothers. He began his own business with a factory at The Gramophone Co.'s premises at Hayes in Middlesex. A flying ground was obtained on what is now Heathrow Airport. Fairey's most famous early aircraft was the Swordfish.

GEC

'Estimates, advice or information given verbally or by letter. Our friends are invited to avail themselves of our aid and to write to us for information upon any subject related to electricity.' Such was the advertisement by the General Electric Apparatus Co. of 5 Great St Thomas Apostle in the City of London which appeared in an electrical catalogue of 1897. The firm was founded by the methodical, realistic and businesslike Mr Gustav Binswanger (he later changed his name to Byng) and the dynamic Mr Hugo Hirst. Mr Hirst was in charge of the so-called 'H' department. It sold dynamos, arc lamps, incandescent lamps, accumulators, cables, ammeters, voltmeters, switches, lamp holders, etc. There was a workshop at Bow and soon the firm acquired works in Manchester. Headquarters were moved to 71 Queen Victoria Street and, in 1921, to Magnet House in Kingsway, the firm now registered as the General Electric Co.

Early on, Hirst opened a lamp factory at Brook Green in Hammersmith previously run by The Brush Co. The company secured the rights to make osmium, tantalum and tungsten filament lamps, and in 1909 made Osram lamps alongside the Robertson Lamp Works. GEC later built research laboratories in Wembley.

George Wailes & Co.

George Wailes was born in Leeds in 1833 and served an apprenticeship with a firm of machine makers in that city before moving to London in 1855. There he took over the premises of Charles Rich, an engineer who was running a small concern in Selby Place, between Gower Street and Hampstead Road. There was a smith's shop, two ground-floor machine shops, iron foundries, a turning shop, a pattern shop, a brass fitting shop, and a boiler and engine house. Wailes was well known for the quality of his castings and by 1889 was employing 148 men. He was an honest man and, because he always paid his suppliers on the spot, was known as 'Ready Money George'. All sorts of machinery were made: Winsor & Newton, the paint makers, were supplied with tube-making machinery; moulds were made for printers' rollers; camera stands for J.H. Dallmeyer; cast-iron violin case moulds, gun boring machines and packing machines for Brooke Bond Tea; and electrical machines, gas engines and parts for Boosey & Hawkes musical instruments.

Many engineering firms dependent on steel or steam power moved out of London in the late nineteenth and early twentieth centuries. Thomas Hancock (rubber processor) moved to Manchester, Donkin's left Bermondsey for Chesterfield in 1902 to make high-speed steam engines, and the chemical engineering firm of George Scott left Silvertown for Fife.

Hydraulic Power

Although patented by Joseph Bramah in 1795, it was Sir William Armstrong in Newcastle upon Tyne who first used hydraulics to power dock machinery. Hull was the first city to have a system of hydraulic mains and London soon followed. Edward Bayzand Ellington, formerly managing director of a hydraulic engineering company in Chester, founded the London Hydraulic Power Co. in the 1880s. The pumping stations were at Falcon Wharf, Bankside; Renforth Pump House at Canada Water in Rotherhithe; City Road Basin on the Regent's Canal; Grosvenor Road, Pimlico; and, most important, Wapping Hydraulic Power Station. Coal-fired boilers raised steam at high pressure which in turn drove hydraulic pumps. The pumps supplied water at an increased pressure to accumulator towers. Water under pressure was then released to a cast-iron ring main system beneath the streets to do its work of opening dock gates, lifting cranes and so on. It was Armstrong who first used accumulators to level out pressure differences between supply (the boilers) and demand (e.g. the lock gates).

The company supplied hydraulic power to dock gates, jiggers, cranes, warehouse lifts and other machinery throughout the docks and in London as a whole via a system of underground mains. In the 1930s, some 8,000 machines were powered through a network of 184 miles of mains, including the safety curtain at the London Palladium theatre. However, the company's activities declined after the Second World War as power from electricity became more commonly used, the stations eventually closing in 1977.

Wapping Hydraulic Power Station.

11 Footwear

While still a young man, John Lobb fell from a donkey – some say from a hay wagon – while working on a farm in his native Cornwall. He sustained a serious injury, but as events turned out, a fortuitous one, for John was forced to abandon farm work and seek employment elsewhere as a boot maker. And in this trade Lobb became skilled; he was also ambitious and, despite his disability, found the strength to walk to London to seek his fortune. At the time, Thomas' of St James's were the best boot makers in London and Lobb headed there and asked for a job. He was met with a firm 'No' and in response retorted that one day, 'I'll build a firm that will knock you sideways'.

Rejected by Thomas', the ambitious young man set off for Australia in search of gold. Gold he did not find; instead he started out making prospectors' boots with hollowed-out heels in which the prospectors could store their gold in secret. Lobb, as we have seen, was never slow at coming forward and while still in Australia was confident enough to display his boots at the 1862 Exhibition in London (a follow-up to the Great Exhibition of 1851). He won a gold medal for 'good work and first class materials' and promptly wrote to the Prince of Wales

John Lobb, St James's Street.

to ask if he could be awarded the royal warrant. His nerve paid off and in 1863 he was awarded it. Three years later, Lobb, his wife and his young apprentice, Frederick Richards, boarded a boat for England.

Despite there being over 3,000 boot makers in London, Lobb thrived. The medals followed: Paris in 1867, Vienna in 1873, and Chicago in 1893. He moved to St James's Street in 1880, with Frederick Richards as manager. Lobb's was badly bombed in the Second World War, forcing a move to 26 St James's Street and later – when *The Economist* building took over the site – to the firm's present address at 7–9 St James's Street.

John Lobb's list of customers is impressive: Sir Thomas Beecham, Enrico Caruso, Cole Porter, Frederick Handley Page, Thomas Sopwith, W. Somerset Maugham, Dennis Wheatley, Amy Johnson, Frank Sinatra, Dean Martin and a host of others.

The firm never embraced machine-operated mass production and always made hand-made shoes. First a last was made, an exact model of the foot. The 'clicker' cut out the leather – the upper could be Russian calf, antelope, doe skin or many other kinds of leather. Next, the 'closer' took over to construct the upper by sewing all the pieces together. The upper then went onto its last and the sole and heel were added.

Boots and shoes were made in large numbers in the East End of London. In 1861, there were 46,000 people employed in the industry. In contrast to Lobb's of St James's, it was a sweated trade with 28 per cent of production in the Shoreditch, Bethnal Green, Stepney, Hackney and Stoke Newington areas. Quality tended to be poor as exemplified by the words of a skilled craftsman giving evidence to the select committee on the sweated system: 'I think there is a very large amount of the population of England that is almost bound to buy the minimum price article of many things.' The industry fell into decline in the twentieth century, with the introduction of factory mass production centred at Northampton.

12 Furniture

High-quality furniture for the wealthy of St James's and Piccadilly was manufactured in the less fashionable parts of Westminster – Covent Garden, St Giles in the Fields and St Martin in the Fields. In this tightly knit area there were highly skilled cabinet makers, upholsterers and looking-glass makers. In the early eighteenth century, in St Martin's Lane, near Long Acre, was the workshop and home of Gerrit Jenson, cabinet maker to the Royal Household in the years from Charles II to Queen Anne. James Moore and his son, of Short's Gardens, St Giles in the Fields, supplied George I, while Benjamin Gordison, at the sign of the 'Golden Spread Eagle' in Long Acre, was cabinet maker to both George II and George III. Most famous of all was Thomas Chippendale, who had a workshop in St Martin's Lane.

Furniture making came to the East End of London in the mid-nineteenth century. London's population was growing exponentially and a market developed for inexpensive furniture. Before, the trade had been centred in the City, in the area between St Paul's Cathedral and Smithfield.

In contrast to the furniture makers of the West End, all of whom employed highly skilled workers trained through formal apprenticeships, the East End trade was concentrated in small workshops. There was a strict division of labour. Frequently a cabinet maker would specialise. For instance, Henry Price of Hoxton only made wardrobes, while the firm of Stubbs of Brick Lane and City Road made chairs. There were also upholsterers, turners and carvers. The final job, before sale, was French polishing. Women of French Huguenot descent expended much elbow grease to apply shellac, a spirit-based polish, to give the wood a smooth and mirror-like finish. French polishers tended to be young women – the 1881 census recorded 248 female French polishers in the East End. Each workshop was a small-scale affair and sometimes an East End furniture maker – perhaps a Jewish or Irish immigrant – merely set up in his own home. In time, trade began to be concentrated in the Curtain Road area. The 1851 census records sixteen furniture makers there, but within ten years many more had joined them, spreading towards Hoxton and Bethnal Green.

Charles Booth, in his *Life and Labour of the People in London*, summed up the trade in Curtain Road:

> From the East End workshops produce goes out of Curtain Road of every description from the richly inlaid cabinet that may be sold for £100 or the carved chair that can be made to pass as rare antique workmanship, down to the gypsy tables that the maker sells for 9/- per dozen or the cheap bedroom suites and duchesse tables that are now flooding the market.

There were, however, some larger manufacturers such as Messrs Clozenberg of Great Eastern Street and B. Cohen & Sons of Curtain Road. The largest was Harris Lebus of Tabernacle Street, at the time the largest manufacturer in the country. They were to move to Tottenham Hale in 1903 and employed over 1,000 people.

Furniture making could be dangerous work. As the old saying went, 'You were never a proper machinist until you had at least two fingers chopped off.' The nearby Mildmay Hospital in Austin Street became known as the cabinet makers' hospital!

The industry boomed between the two world wars, responding to London's expanding housing stock. But by this time mechanisation had become established and firms, larger in size, moved out to the Lea Valley, so beginning the decline of the traditional Curtain Road workshop.

13 Glass

Southwark, and Bankside in particular, were well known as centres of the glass industry in London. Roman glass has been found there, and German-born Bernard Flower, royal glazier to Henry VIII, had lived just to the east of Old London Bridge and had led a group of skilled craftsmen. Their handiwork could be found in Old St Paul's Cathedral and the Tower of London. But in the Middle Ages, glassmaking was a rural industry concentrated in the Weald of Sussex and Kent.

Glass is made by fusing sand and limestone, with sodium carbonate added as an alkaline flux to lower the melting point. In medieval times wood ash provided the alkaline flux, with wood also used as fuel for the furnace. Wood was a valuable commodity, used extensively in shipbuilding, and this prompted James I, in 1615, to issue a royal proclamation forbidding its use as fuel in glassmaking. Coal had to be employed instead and this posed real problems for the ancient Wealden glass industry. But the Weald's problem was Southwark's opportunity. Coal could be brought in easily on the Thames and there was a ready market for the finished product in the City of London.

Vice Admiral Sir Robert Mansel had a monopoly on glassmaking at the time with his patent for making 'all sorts of glass with pit coal'. He was prevented from building glasshouses in the City because the smoke from coal burning had always been opposed by the clergy and City Fathers – a man was hanged for burning coal in the thirteenth century! – but regulation south of the river was lax and so Bankside was an obvious choice. Accordingly, in 1613, the first coal-fired furnace began production in the precincts of Winchester Palace in a building previously used as a brewhouse. Glassmaking continued for many years at this site until manufacture stopped when the area known as Glasshouse Yard was cleared for the construction of the Borough fruit and vegetable market.

There were other glasshouses in the vicinity of St Mary Overy Church (now Southwark Cathedral) which could have been in operation as early as Tudor times making window glass. In the 1670s, they were owned by Captain Thomas Morris and glass bottles were made there. These were made using the ancient skill of glassblowing, introduced by the Romans. A metal pipe was dipped into molten glass and a quantity collected on its end. This was blown into a mould of preformed shape. Bottle glass was usually of poor quality and often coloured green because of iron impurity.

John Bowles, born in Chatham in 1640, was a well-known Bankside glassmaker. He had works, opened in 1678, at the junction of Deadman's Place (now Park Street) and Stoney Street and other glasshouses in the Bear Garden, close to the site of Shakespeare's Rose Theatre. It was at the Bear Garden glasshouse that Bowles, in 1691, pioneered the use of Crown glass for the manufacture of window panes.

Until the mid-nineteenth century most window glass was made by Bowles' Crown method. The skilled glassblower would blow a hollow glass globe on a lump of glass attached to a pontil (a metal rod). The glass was then reheated and the pontil rapidly spun so that the glass collapsed into a flat disc by centrifugal force. After annealing, panes of glass were cut from the disc. The blemish in the centre where the pontil was attached became the 'bull's-eye' pane often seen in old windows. Window-glass panes were small in size; the largest that could be made were only 24in by 15in, which explains why the panes in old buildings are so small. The name of Crown glass comes from Bowles' trademark – a crown which he embossed in the centre of each pane. Bowles was soon to move his operations north of the river to Ratcliffe, his Bear Garden glasshouse being taken over by a syndicate which also made window glass. They obtained a patent for 'casting glass and particularly looking-glass plates, much larger than was ever blown in England or any foreign parts'. Bowles' Ratcliffe works, the Cockhill Glass Works in Glasshouse Fields, were situated near to the present-day entrance to the Limehouse Link and by 1731 consisted of two glasshouses, three warehouses and the manager's house.

Glass was also made in Lambeth. In 1615, Sir Edward Zouch, no doubt responding to James I's proclamation of that year, obtained a patent for making drinking-glass vessels and opened a factory in Lambeth. It was taken over, in the 1660s, by the Duke of Buckingham and then by John Bowles in partnership with a Mr Dawson, where 'it excelled the Venetians and other nations in brown glass plate'. The works was in Vauxhall Square, about 200yd north of present-day Vauxhall Bridge. Glasshouse Walk is still there today to remind us of the two glasshouses which made a 'prodigious profit' – the Great Glasshouse was insured at a value of £1,000 and the Little Glasshouse at £400.

There were glasshouses on the site of what is now Tate Modern in Bankside. These were run by the Jackson family, well-known glassmakers who also had works in King's Lynn. It was Francis Jackson, born in Bridgnorth in 1659, who built the first cone glasshouse in Bankside. These were large brick-built structures, 80ft in height and 40ft across. The furnace was in the centre and the glassmakers worked in the annular space within the cone in teams of men known as 'chairs'. The team leader was known as the 'gaffer', the first use of this now well-known word.

Lead crystal or flint glass was introduced to London by George Ravenscroft in 1675, and it was soon made in Bankside. Crushed flint and lead oxide were used in place of lime, the finely worked glass being used for domestic ware. In 1693, at their Falcon Glassworks to the east of Gravel Lane, Francis Jackson and John Straw were making 'the best and finest drinking glasses and curious glasses for ornament'. But they were not universally popular. Bankside women took in laundry from rich merchants across the river and did not take kindly to soot from the glasshouses falling on their clean washing. They even petitioned the King in 1688 that 'one John Straw and others are erecting glasshouses in the middle of the parish to the utter ruin of many of the inhabitants whose livelihood depends upon washing'. The Jackson dynasty came to an end in Bankside in 1752 and there was then a succession of owners at Falcon Stairs, including Pellatt & Green

Elevation of the furnaces, and interior view of the Glass-house
and working operations.

Glassmaking at Apsley Pellatt's glassworks, Southwark. *(By permission of Southwark Local History Library)*

who later opened other works at Upper Ground, in the grounds of Paris Manor, a little to the west. Pellatt & Green, 'Glass Makers to the King', became famous the world over for their superb cut glass and cameo encrustation. Michael Faraday carried out much research into optical glasses at their works, such was the quality of their glass.

Apsley Pellatt IV was the most eminent member of the family. He entered the Falcon works in 1810, and during the following years took out many patents on improved methods of glassmaking, as well as publishing books. Pellatt's cameo encrustations were medallions or other ornaments enclosed in glass and, in his own words, 'like the fly in amber they effectively resist for ages the destructive action of the atmosphere'. Highly suitable as commemorative records in the foundation stones of new buildings, they have been found as far afield as Philadelphia. Pellatt's inscribed slabs were placed in Windsor Castle at the time of Wyatville's restoration, in the Wellington Barracks at the Tower of London, and in the Royal Exchange. The firm moved north of the Thames at the end of the nineteenth century, thus bringing glassmaking in Bankside to a close.

The first glassworks in the City of London had been established in 1575 by a Venetian, Jacob Verzelini, at Crutched Friars in Aldgate – despite the City Fathers' objections. Later, in 1680, William Davis built himself a glassworks on the site of the demolished Whitefriars monastery between the Temple and the River Fleet. It was an ideal site as sand imported from France and coal from Newcastle could be landed at the nearby river wharf. Until the mid-nineteenth century it made flint glass. An advertisement in *The Tatler* of 1710 tells us that 'at the Flint Glass House in White Fryars near the Temple are made and sold by wholesale or retale, all sorts of decanthars, drinking glasses, crewits etc or glasses made to any pattern of the best flint'. The glassworks passed through various hands before being taken over by James Powell in 1834. It was at this time that the medieval art of staining glass was in danger of being lost forever. But Charles Winston, archaeologist and barrister, came to the rescue by collecting samples of medieval glass and having them chemically analysed. He gave the results to Harry Powell, the founder's grandson, who managed to produce glass with colour and translucency comparable to medieval work. So was born James Powell & Sons' Whitefriars stained-glass studios. Many eminent nineteenth-century artists designed for the company, including Dante Gabriel Rossetti, Ford Madox Brown, William Morris, Edward Burne-Jones and Sir Edward Poynter. Designers would produce sketches, a cartoon would be made and then a cutline. Pieces were then cut by the glazier and given to the painter. In 1919, Powells moved to Wealdstone. On their first day of operation, live coals from Whitefriars were carried there to keep alive the tradition that the furnaces had fired continuously since 1680.

Other glasshouses on the north side of the river belonged to Mr Lewis in the Minories, and to Mr Dallow in Rosemary Lane, Whitechapel. The Thames Plate Glass Co. of Bow Creek, Poplar, made plate glass for clients far and wide. All the glass for the Crystal Palace of the 1851 Great Exhibition was made here. This specialist firm finally closed in 1875.

14 Leather

The exact date when the leather industry began in Bermondsey is unclear. Chaucer in the fourteenth century refers to the 'tan yards of Bermondsie', and in the Elizabethan era there were tanneries in the vicinity of Long Lane.

It is easy to forget that Bermondsey was once the centre of London's leather trade. Local street names like Tanner Street, Morocco Street and Leathermarket Street remind us of this lost industry. And the stench that permeated from the dripping skins of dead animals was abominable – it would have made the 'Great Stink' that emanated from the Thames in the mid-nineteenth century seem like the perfume department at Selfridges.

Bermondsey was well placed to be the centre of the leather industry. As long ago as 1392, a proclamation decreed that butchers in the City should dump their animal skins on the other side of the river. The Thames also provided a plentiful supply of water for the tanning process and there was a ready market in the City for the finished product. At one time there were over fifty firms engaged in the trade.

Hides arrived in Bermondsey from local slaughter houses, and were also imported from abroad. They were salted before dispatch by soaking in brine to prevent deterioration. There are many different sorts of hide, each suitable for a different end product. The buffalo has a very thick and tough hide which is used for industrial purposes. In contrast, calf skin is much thinner and more supple and is used for luxury articles such as handbags and purses. Pig skin has a compact grain and is used for articles that are subject to friction, such as pocket wallets and saddles. Cattle hides have the greatest versatility and are used for shoes, upholstery and industrial belting. Animal hides have three layers. The outer is the epidermis, a layer of cells with follicles containing hair or wool. The second layer is the corium. It is the most important part of the hide and is used to make leather. The inner layer is a fleshy substance that connects the corium to the flesh of the animal.

There are a multitude of different trades between the slaughterhouse and the finished article: skinners, fellmongers (who separate wool and hair from the pelt), tanners, curriers (who shave, stretch and grease the leather after tanning), dyers and whitetawyers or greytawyers (who taw skins using alum). Here are details of the process. Anyone with a squeamish nature may want to bypass this section! After the poor beast was slaughtered, its flesh was sold to the butcher and the hide to the tanner. The horns were first removed and sold to comb makers. Then it was the job of the 'flesher' to scrape the hide to remove small pieces of flesh. The hide was then 'unhaired' by soaking in a quicklime solution for several days, after which the roots were sufficiently loosened for the hair to be easily pulled out. 'Graining' was the next job whereby the hide was soaked in a dilute solution of

Barrow, Hepburn & Gale, Long Lane, Southwark.

Bevington's Tannery, Bermondsey. *(By permission of Southwark Local History Library)*

Breaking, unhairing and fleshing sheep skins, Bevington's Tannery, Bermondsey. *(By permission of Southwark Local History Library)*

Relief at the Leather Exchange, Southwark, showing female workers fleshing a hide.

sulphuric acid to open the pores. Following this, the pelt, as it was then known, was left to soak in tanning pits containing a solution of oak bark. The solution, appropriately known as 'ooze', seeped into the pores from surface to surface. After several months in the tanning pit, the leather, as it had become, was removed, hung up to dry and finally pressed between rollers. It was then off to the shoe-maker for the best pair of boots a City gent could wish for.

There were scores of Bermondsey firms involved in the leather industry. Perhaps the best known were Barrow, Hepburn & Gale and Bevington's. Barrow, Hepburn & Gale Ltd were based at The Grange Tannery and Bermox Tannery. A tannery at The Grange is shown in a map of 1750, and in 1829 a Mr Alexander Ross had a leather works in Grange Road. He was joined by a Mr Tomlin from Wellingborough who later became a proprietor. In 1901, Hepburn & Gale built tanneries in Grange Road and two years later combined to form Hepburn, Gale & Ross Ltd. They later merged with Samuel Barrow to form Barrow, Hepburn & Gale.

In the late nineteenth century, the leading man in the leather industry was Samuel Bourne Bevington. He was Bermondsey's first mayor, between 1900 and 1902, and was known locally as 'the Colonel', a reference to his military back-ground as commanding officer of the 3rd Volunteer Battalion of the Queen's Royal West Surrey Regiment. He did much work for charity and it is said that he was a 'straightforward man whose diction of short jerky sentences always sug-gested the soldier'.

The Bevingtons have been in the tanning business for centuries. In 1660, William Bevington was working as a tanner in Warwickshire. His son, John, was also a tanner. In the mid-eighteenth century John's son, Samuel, travelled to London and was apprenticed to a Mr Peck, a Master of the Leathersellers'

Company. Samuel was joined in the trade by his four sons. They set up at 34 Gracechurch Street and purchased lamb and kid skins from Italy which they sold to dressers in Yeovil to make gloves; these were then returned to London for sale. There were members of the Bevington family living in Yeovil and, in 1800, Samuel Bevington, who had been working as a leather dresser and glove maker at Stoke-under-Ham in Somerset, came to London and began making leather in Grange Road. He was joined by his brother, Henry, and the pair opened at Neckinger Mills in 1806 on the site previously occupied by a paper factory.

By the 1980s, the leather trade had left Bermondsey. All that remains are the old premises, now converted to luxury apartments or offices.

15 Match Making

Match making was known to the Chinese. A forerunner of the modern match – a pinewood stick impregnated with sulphur – was invented in China by ladies of the Northern Qi Court in AD 577. They used them to light fires to cook food. Later, a Chinese author wrote:

> If there occurs an emergency at night it may take some time to make a light or light a lamp. But an ingenious man devised the system of impregnating little sticks of pinewood with sulphur and storing them ready for use. At the slightest touch they burst into flame. One gets a little flame like an ear of corn. This marvellous thing was formerly called a light bringing slave but afterwards when it became an article of commerce its name was changed to fire inch stick.

In Britain, Robert Boyle and Godfrey Haukwitz experimented by coating wooden splints with phosphorus and sulphur. However, it was not until 1827 that the first practical 'friction' match was invented by John Walker, a 'chymist and druggist' from Stockton-on-Tees. He mixed together potassium chlorate, antimony sulphide and gum arabic, and found that when coated onto splints and drawn across a rough surface the mixture burst into flames. Within three years, Walker had increased his production from 1,836 tins to over 12,000 sold every year; they were very popular with smokers and young men who enjoyed throwing the lighted splints at ladies' feet to give them a shock! Walker's matches gave off an offensive smell and phosphorus was later added to remove the odour. They were known as Lucifers (the bearer of light) or Congreves (after Sir William Congreve, the rocket inventor). Unfortunately, white or yellow phosphorus is poisonous and it was not unknown for babies to be poisoned by swallowing them. Samuel Jones sold matches in the Strand and warned his customers: 'If possible avoid inhaling the gas that escapes from combustion of the black composition. Persons whose lungs are delicate should by no means use the Lucifer.' Early match making was a cottage industry and workers – frequently children – worked long hours for little wages. Their health suffered and many developed phossy jaw, an extremely unpleasant disease which causes the jawbone to collapse and glow green in the dark. As an eyewitness described one such afflicted, 'You could take his chin and shove it all in his mouth.'

William Bryant was born in Tiverton in 1804 and, together with Edward James, ran a general merchant's shop in Plymouth. They also started a soap factory and took out a patent for 'Improvements in the Manufacture of Liquid and Paste Blacking by the Introduction of Indian Rubber, Oil and other articles and Things'. Meanwhile, Francis May had opened a grocer's shop and tea dealership at 20 Bishopsgate. He became the sole agent in London for Bryant and James's patented 'India Rubber and Blacking'. In 1843, Bryant and May – both Quakers – joined forces to found the provisions merchants Bryant & May, at 133–4 Tooley Street in Southwark.

Bryant & May

The market leaders in match production at that time were the Swedish firm run by Carl and Johan Lundström in Jönköping. They were keen to export matches to England and so Carl Lundström journeyed to London to look for firms to import and sell his matches. After a number of fruitless meetings he was recommended to visit Bryant & May. They ordered some of Lundström's matches and eventually became the sole agents for the safety matches which the Swedish firm had patented in 1855. The safety aspect came from separating the combustible materials on the match head from the special striking surface, as well as replacing toxic white phosphorus with red phosphorus.

The striking surface of the safety matches was composed of 25 per cent powdered glass, 50 per cent red phosphorus, 5 per cent neutraliser (calcium carbonate), 4 per cent carbon black and 16 per cent binder. The match head consisted of 45–55 per cent potassium chlorate, sulphur, starch and antimony sulphide, with the remainder being siliceous filler and glue. Demand in Britain was phenomenal – in 1851, Bryant & May imported over 1 million boxes; by 1860, the figure was nearer to 30 million. But the plain fact was that the Swedish firm just could not keep up with demand, so Bryant & May took the obvious step – with the help and guidance of Lundström – of making matches themselves. They opened a factory on a 3-acre site at Fairfield Road, Bow, previously occupied by a crinoline maker and a rope works.

The manufacturing process consisted of taking splints, twice the length of the eventual match, and 'filling' 3,900 of them into a special frame. They were then immersed in paraffin to aid combustion and each end dipped into the igniting mixture. After drying the matches would be 'racked out' of the frame and put on wooden trays where the 'cutters down' hand cut them in half, thus producing 7,800 matches which the 'packers' would 'count in' to each box.

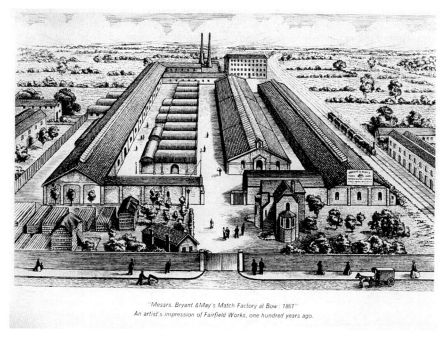

"Messrs. Bryant &May's Match Factory at Bow: 1861"
An artist's impression of Fairfield Works, one hundred years ago.

Bryant & May's Fairfield match factory, 1861. *(By permission of Tower Hamlets Local History Library)*

Match-makers in the East End, 1871

Match girls at Bryant & May, 1871. *(By permission of Tower Hamlets Local History Library)*

All was well in the match-making industry until 1871, when Chancellor of the Exchequer Robert Lowe imposed a match tax of ½d on every box. The match girls saw this as a threat to their livelihood and 3,000 of them assembled at a mass meeting in Victoria Park to plan what to do. They decided to march on Parliament and submit a petition. Despite scuffles with the police at Globe Bridge and Mile End, they eventually arrived at the Embankment at Blackfriars from where a small number proceeded to Westminster Hall to hand in the petition. Meanwhile, back at Blackfriars a riot had broken out; police threw the girls' placards into the Thames and Lowe had his way to the House of Commons blocked, obliging him to enter via an underground tunnel. But the girls' persistence paid off, for soon Lowe was to tell the Commons: 'I have announced that the measure with regard to matches will not be proceeded with.' By way of celebration, Bryant & May erected a drinking fountain (now demolished) in Bow and with stunning thoughtfulness allegedly docked 1s from each girl's wages to pay for it, a course of action that the girls were not to forget.

Life as a match girl was repetitive, hard work, dangerous (Bryant & May still used white phosphorus for some matches because of their lower cost and greater demand) and the hours were long. Their plight came to the attention of Annie Besant, the radical socialist activist, while she was listening to a lecture in June 1888 at a Fabian Society meeting. Next day she went to the factory to speak to the girls and was told the women worked for fourteen hours a day for less than 5s a week and that they were often subjected to fines by the Bryant & May management for offences such as talking, going to the toilet without permission, dropping matches or being late. Later in the same month, she wrote in her newspaper *The Link* an article entitled 'White Slavery in England', contrasting the conditions the girls worked under with the very handsome dividends paid to Bryant & May shareholders:

> Born in slums, driven to work while still children, undersized because under-fed, oppressed because helpless, flung aside as soon as worked out, who cares if they die or go on the streets provided only that Bryant and May shareholders get their 23% and Mr Theodore Bryant can erect statues and buy parks?

Bryant & May responded by trying to persuade the girls to write a letter refuting Besant's claims. They refused, the organisers were promptly sacked and all 1,400 women walked out. *The Times* newspaper was less than sympathetic, blaming left-wing agitators:

> The pity is that the matchgirls have not been suffered to take their own course but have been egged on to strike by irresponsible advisers. No effort has been spared by those pests of the modern industrialised world to bring this quarrel to a head.

Frederick Bryant threatened to sue Besant but she proved a tough and formidable opponent. She took fifty match girls to Parliament to lobby a group of MPs

Match making at Bryant & May in the 1970s. *(By permission of Tower Hamlets Local History Library)*

and later they linked arms and marched along the Embankment. The girls stayed out for three weeks until the London Trades Council arranged a meeting with Bryant & May, which eventually conceded most of the girls' demands, in particular by agreeing to abolish fines. In late July 1888, the Union of Match Makers was founded with Annie Besant as its first secretary.

A major advance in match making came in 1898 when two French researchers discovered phosphorus sesquisulphide – a compound that can produce 'strike anywhere' matches. Bryant & May purchased the patent rights and Albright &

Wilson began supplying the new material. Its introduction had an added bless-ing – it heralded the end of phossy jaw and later the Berne Convention of 1906 banned the use of yellow or white phosphorus altogether.

By the early twentieth century a rival company, the Diamond Match Corporation of America, had bought a factory in Bootle and began making the soon-to-be well-known brands of 'Captain Webb', 'Puck' and 'Swan Vestas'. They were cheaper than Bryant & May, prompting the two companies to merge in 1901. Bryant & May then built a new factory at Bow, modelled on the Diamond Works in Liverpool.

By 1911, 2,000 women were working there at what was then the largest factory in London. The company acquired S.J. Moreland of Gloucester ('England's Glory') and, in 1926, combined with others to form The British Match Corporation, which merged with Wilkinson Sword in 1973. In the early 1970s, the decision was taken to discontinue match making at Bow, and the works closed in 1979. The factory remains as Bow Quarter, a complex of luxury apartments.

There were smaller-scale match makers in the Stratford area. Bell & Black were in the High Street between 1839 and 1882, Benjamin Daniels was in Martin Street between 1886 and 1905, and G.M. Judd & Bros were in Carpenters Road between 1908 and 1927.

16 Paper, Printing, Newspapers & Bank Notes

Paper

The first mention of paper making in Britain was in 1450 when Tate's Mill was recorded as operating in Stevenage. Paper was made from rags; the best quality paper from white linen rags. From the latter half of the nineteenth century cel-lulose fibre from wood was increasingly used. Today, although machine-operated, the principles of manufacture remain more or less the same.

Rags were soaked in water and left to ferment for two or three months. They were then cut into pieces and placed in a trough which was continuously fed with water. The rags were then mashed into a pulp by a water-powered stamp with cutting edges. In 1650, Dutch paper makers developed the Hollander, a wooden roller consisting of a cylinder of knives. The roller was rotated in a cov-ered drum-like tub, powered by a water wheel. In the tub were the water-soaked rags ready for disintegration.

When a fine pulp had been formed, the vatman plunged a wooden mould into the vat of pulp. The mould consisted of a square frame covered with a fine wire mesh and resting on another frame, the deckle. The skill of the vatman was to obtain a uniform layer of pulp. He then drained off the water before handing the mould to the couchman who took off the deckle and placed the sheet of pulp (now becoming paper) onto a piece of felt. The felt and paper were laid as alternate layers and when a sufficient height was reached both were pressed with a screw press. The felt and paper were then separated; the paper was pressed once more and then hung up to dry.

The disadvantage of this early way of making paper was that it was a batch process, but in the early years of the nineteenth century Henry Fourdrinier (1766–1854), assisted by Bryan Donkin, developed a machine to make continuous rolls of paper. Pulp was fed onto a moving belt of gauze and the water was squeezed out of it. Later, because of a shortage of raw materials, wood pulp and esparto grass were used as substitutes for rags. The esparto grass was imported from Spain.

Early paper mills in England were at King's Langley and Dartford, and in London at Wandsworth along the banks of the River Wandle.

Wandsworth Paper Mills

In 1836, Thomas Creswick opened a paper mill in Garratt Lane, on the banks of the River Wandle, on a site previously occupied by the Henckels ironworks and before that by a copper works. He styled himself 'Papermaker and Card Maker to His Majesty' and as well as paper, manufactured playing cards and drawing paper. In the succeeding years the mills were owned by a number of concerns, such as Easton & Amos – better known in Southwark as manufacturers of steam engines. They ran a paper mill from 1849 to 1852. Their mills had four beating engines to pulverise the rags, powered by a 16hp water wheel.

Fifteen or so miles to the south-west were William MacMurray's Royal Mills at Esher, with thirteen beating engines, the largest in Surrey. In December 1853, they were destroyed by fire and MacMurray, rather than rebuild at Esher, took over the vacant site of the Wandsworth Paper Mills and installed his brother, James, as supervisor. As well as manufacturing paper, MacMurray supplied newsprint for *The Times* and the *Illustrated London News*. He also owned land in Spain and North Africa from where he imported esparto grass to a dock at the mouth of the Wandle known as MacMurray's Canal (see Chapter 21). Poor William MacMurray was certified as a lunatic in 1883, soon after changing his will to exclude his brother. He died in 1888, but James successfully challenged the will.

The mill fell victim to a serious fire in 1903 when many of its buildings were destroyed. It was then the largest paper mill in London and employed about 160 people. It opened again, but not for long; it was demolished in 1910 and the site was taken over by a varnish works, a gas mantle factory and an engineering works. Esparto Street is a present-day reminder of Wandsworth's paper-making past.

There was another paper mill, the Wandle Patent Pulp & Paper Co., situated on the banks of the Wandle just south-west of Earlsfield railway station in the mid-nineteenth century. By 1873 it had become a bone mill.

Lepard & Smith

William Lepard, son of a Baptist bricklayer, started in the paper trade in 1757 at – as his business card said – 'the second house from the bridge foot in Tooley Street, Southwark', where 'he sells all sorts of stationery wares, paper hangings, Bibles, Testaments, spelling books, etc., wholesale and retail. The best prices given for all sorts of Rags for making paper.' His account books inform us that he was frequently paid in kind: his barber in Tooley Street gave him free shaves for six months to pay for 15s worth of stationery. In time, William Lepard acquired Hamper Mill, on the River Colne near Watford, and was joined by his son-in-law, John James Smith, and cousin, John, the firm now becoming Lepard, Smith & Lepard. From here a horse-drawn van delivered paper to the firm's premises now in London's Seven Dials.

In the early nineteenth century, activities associated with the paper industry were subjected to a series of stringent taxes. At Lepard & Smith, 'every ream of paper had to be weighed and stamped by Government officials and all windows and doors had to be sealed with red wax every night. There were rigorous penalties for anyone falling foul of Excise Officials.'

But then the paper and printing industry got an enormous boost with three Acts of Parliament in the mid-nineteenth century. First, in 1854, William Gladstone removed the advertisement duty; then, one year later, the newspaper stamp duty was abolished; and finally, in 1861, the paper duty was done away with. The consumption of paper exploded: in twenty years it increased two and a half times. Now, 'new publications sprang up like mushrooms. Printing establishments worked day and night turning out new periodicals and cheap reprints of existing books and setting up fresh ones as fast as the authors could write them.'

A fire at Hamper Mill in 1878 forced its closure and Lepard & Smith ceased making paper (the removal of duty having enabled cheap foreign paper to be imported), and concentrated on operating as a print house. In 1907, they moved to Earlham Street, Covent Garden.

There were paper mills on the banks of the River Lea. In the 1840s, a Mr Petrushkin set up a paper mill using hemp which he obtained at the docks from unwanted rope. He was thus able to manufacture a better quality paper and hence obtain 'money for old rope'! Also in Bow was Edward Lloyd, who in the 1860s used first straw and then esparto grass. He employed about 200 hands, working two twelve-hour shifts and getting paid between 10s and 13s a week. There was a branch of Wiggins Teape at Bromley-by-Bow whose mill also used esparto grass. Also on the banks of the Lea at Wick Lane was the wallpaper business of John Allan. At the end of the nineteenth century he sold it to the Wall Paper Manufacturers Ltd, which became a thriving concern employing over 200 people. In the 1960s, it became part of the Reed Paper Group.

Printing

Credit for the invention of printing goes to the Chinese, but printing as we know it today was pioneered in the mid-fifteenth century in Germany by Johannes Gutenberg. William Caxton brought printing to London. He was born in the Weald of Kent in about 1422, and in 1438 served an apprenticeship under the wealthy London mercer Robert Large. Caxton left for the Continent in 1441, and in 1468 was working as a translator and secretary to Margaret, Duchess of Burgundy, the sister of Edward IV. According to one version of events, Caxton learnt the craft of printing from Colard Mansion in Bruges and such was the burden of all the handwriting he was required to do for the duchess, he turned to printing: 'Therfor I have prectysed and lerned at my grete charge and dispense to ordeyne this said book in prynte after the manner and forme as ye may here see.'

In 1476, Caxton returned to London and set up shop near Westminster Abbey at the sign of the 'Red Pale'. His first book – and the first book printed in England – was *The Dictes or Sayengis of the Philosophres*. Within three years he had printed thirty others. Wynkyn de Worde was Caxton's assistant, both in London and Bruges. He disputed that Caxton learnt his trade in Bruges and maintained it was in Cologne where he learnt to print: 'And also of your charyte call to remembrance that soule of William Caxton, the first printer of this boke, in later tonge at Coleyn, himself to avaunce that evefry well disposyd man may theron loke.' After Caxton died in 1491, Wynkyn de Worde remained for a while at the 'Red Pale' until moving in 1500 to the 'Sunne' in Fleet Street, so beginning that street's long association with the printing trade.

Others followed Caxton's lead. John Lettou and William Machlinia printed law books 'in Holborn' and 'by Flete brigge'. Their business was taken over by the Normandy-born Richard Pynson, who had trained in Rouen and was described as 'the finest printer this country had yet seen'. He moved to 'The George' in Fleet Street, close to the church of St Dunstan in the West. Pynson took over from William Facques (of Abchurch Lane) as the royal printer at a salary of 46s per year, which was later raised to £4, and printed royal proclamations and government statutes.

John Day was a celebrated printer. He was born in Suffolk in 1522, and by 1546 was in partnership with William Seres in Snow Hill. Day left Seres in 1551 and moved to Aldersgate. As we shall shortly see, printing could be a dangerous profession and in the reign of Queen Mary, Day fell foul of the law, was arrested and was clapped in the Tower 'for pryntyng noythy [naughty] bokes'. Meanwhile, the Worshipful Company of Stationers was incorporated by Royal Charter to regulate the trade and enable the government to keep a firm check on all printed matter. All books had to be printed with the permission of a 'Special Commission of Ecclesiastical Authorities' on pain of three months' imprisonment. When Queen Elizabeth I came to the throne, John Day found a useful patron in Archbishop Matthew 'Nosy' Parker and at Lambeth Palace he printed for Parker *De Antiquitate Britannicae Ecclesiae*. He also printed Asser's *Life of King*

Alfred and Foxe's *Book of Martyrs*, a copy of which Elizabeth decreed should be in every parish church in the land.

Twenty years after Day's incarceration in the Tower of London a book came to light defending political prisoners. It was published by Thomas Cartwright and printed in secret by a man called Asplyn who turned out to be one of Day's apprentices. Forgetful that he, himself, had been a victim of government censorship and a political prisoner, it was Day who discovered the offending press and alerted the authorities. Asplyn wanted revenge – he attempted to murder Day and his wife. Day became Master of the Stationers' Company in 1580. He was twice married and had twenty-six children.

There was vigorous censorship in the reigns of James I and Charles I, and many printers and authors were familiar with the inside of jails, if not worse. William Prynne was a barrister at Lincoln's Inn. He wrote a book criticising the theatre and actors. In it he referred to the presence of ladies at stage plays. The authorities assumed he was attacking the Queen who was in the habit of attending masques. Poor Prynne had part of his ears cut off and was fined £1,000. It did not deter him and he continued to write, for which the remainder of his ears were sliced off and he was branded with S.L. on the cheek: 'Scurrilous Libeller'. In similar vein, Robert Barker printed a copy of the Authorised Version of the Bible but omitted to include the word 'not' in the seventh commandment – thus in all innocence he accidentally encouraged his readers to commit adultery. The book became known as the 'Wicked Bible' and for his error Barker was fined the huge sum of £300. Later in 1666, John Twyn published a pamphlet, *A Treatise of the Execution of Justice*, in which it was alleged that he was inciting rebellion and encouraging people to kill the King. His apprentice, Joseph Walker, was a witness for the prosecution; Twyn was found guilty and was executed at Tyburn.

In 1636, an Act of Parliament formalised restrictions placed on authors, printers and publishers. All books had to be licensed: law books by the Lord Chief Justice; history books by senior Secretaries of State; books on heraldry by the Earl Marshal; and everything else by either the Archbishop of Canterbury, the Bishop of London or the Chancellors of Oxford or Cambridge.

Throughout the seventeenth and eighteenth centuries, printing techniques showed little change from Caxton's day. Letterpress printing consisted of applying ink to letters – formed back to front – and standing in relief. The paper was placed on this surface and then pressure exerted. An alternative method was intaglio printing, whereby ink was applied to engraved hollows, the surface wiped clean, paper put in place and pressure applied as before. There were other trades associated with the printing industry – typesetters and type founders. William Caslon was London's best-known type founder. He was born in Cradley and began in London as a gunsmith before setting up a type foundry in 1716.

There were technical advances at the beginning of the nineteenth century. In 1798, a German, Alois Senefelder, invented lithography. It depends on the chemical principle that oil and water repel each other. He applied the material to be printed as an oily substance to a smooth surface of stone. Ink, applied by a roller, would be attracted to the oily surface but repelled by the surface left free of oil. By this means he was able to get an impression of the inked surface when pressure was applied.

Until the end of the eighteenth century, all printing presses were made of wood and were operated manually by pressmen. Then, Charles, the 3rd Earl Stanhope, in collaboration with Robert Walker of Dean Street, Soho, set about increasing the speed of printing by pioneering a press, the Stanhope Press, built entirely of iron. It relied on a series of levers to compound the pressure, making life easier – if only slightly – for the pressman. It was quickly adopted by the newspaper industry. Then, in 1806, Frederick Koenig arrived in England from Saxony. He built the first steam-powered cylinder press which was quickly taken up by *The Times* newspaper, which in 1814 produced the first newspapers printed with the aid of steam power. The first machine-printed issue was produced without the knowledge of the hand-press men. When they found out they were warned not to attempt violence because 'there was a force ready to suppress it'. The machine could produce 1,800 copies per hour – six times as many as by the hand press, T.C. Hansard commenting that 'the great object in the employment of machinery is to lessen the expense of printing'.

The House of Harrild

R. Harrild and E. Billing began producing broadsheets (early newspapers) at the Bluecoat-Boy Printing Office in Russell Street, Bermondsey, in 1801. The two men were brothers-in-law but by 1807 parted company, leaving Robert Harrild at 127 Bermondsey Street as a printer and publisher of books. He moved to 20 Eastcheap in the City in 1819 and began specialising in children's books which had what could be described as a 'moral uplift'. Titles included *The Youth's Guide to Wisdom*. There were educational books as well, such as *Juvenile Cabinet of Natural History*. In the 1820s, Harrild began to diversify into commercial printing and expanded into new premises in Friday Street and nearby Distaff Lane. Ten years later, he gave up printing altogether and began making printing presses – advertising as 'Printers' Composition Roller Maker, Appraiser, Joiner, Manufacturer of New and Dealer in all kinds of Second Hand Printing Materials'.

Harrild's company used rollers to ink their printing plates and supplied their machines to *The Morning Post*, *The Globe*, *Shipping Gazette* and *The Weekly Chronicle*. They exhibited at the Great Exhibition of 1851 and one year later moved to larger premises at 25 Farringdon Street. The firm developed a ten-feeder press in which ten men fed paper at the same time. Ten rollers inked the type which was on a cylinder at the centre of the machine. It was bought by *The Times* newspaper in 1861 which was then able to produce 1,000 sheets every hour. Advances made later in the century increased the run to 40,000 sheets per hour.

Many publishers used Harrild's machines, including the *Illustrated London News*, *The Graphic*, *The Field* and *Lloyd's Register*. New premises were acquired between Fetter Lane and Furnival Street, but it proved to be a bad move and in 1905 Harrild's moved to Stamford Street in Southwark.

W. Lethaby & Co. Ltd

The firm was founded by W. Lethaby. He was born in Devon in 1875 and moved to London in 1890 to be apprenticed as an engraver. By 1913, he had been joined by Richard Davis. The firm adopted the trademark LEDA and made numbering machines. They were based at 12 Hatton Wall (off Hatton Garden) and in 1916 moved to 124–126 Clerkenwell Road, making all sorts of numbering machines for order forms, invoices, tickets, bills and receipts for the printing industry.

Printing in Southwark

There has been a long and fine tradition of printing in Bankside. Peter Treveris printed his *Grete Herball* in 1529 and was described as 'this ingenious and elegant printer'. Max & Co. of Borough Road were pioneer lithographic printers. Their proud boast was that they were the first to use steam to drive their presses. Nearby, in Southwark Street, Barclay & Fry printed cheques and business stationery. Burrups opened at the turn of the nineteenth century in Southwark Street. They had started 'At the Sign of the Crane' in the City in 1628 and began the custom of using the endpapers of their books to advertise other titles. John and William Burrup were keen followers of cricket. Associated with Surrey County Cricket Club, William took the first England touring side to Australia in 1861, long before proper test matches or the Ashes were thought of. Burrups' reputation rested on their work for the City of London: each morning they had an advert at the top left-hand corner of the *Financial Times*, promising 'Speed, Accuracy and Security'. During the General Strike of 1926, the directors of Burrups were forced to work the presses themselves; they printed the *Daily Express* which, at the time, was the only national newspaper to announce the birth of Princess Elizabeth, now Queen Elizabeth II. As well as their work for the City, they also printed *Boy's Own Paper* and saucy seaside postcards. Burrups were bombed in the Second World War and were forced to move from Southwark for a while. The firm returned in 1959 to premises in Lavington Street and in 1964 were taken over by the Exchange Telegraph Co. St Stephen's Parliamentary Press, part of HMSO, opened in 1961 in the presence of the Lord Chancellor and the Speaker of the House of Commons. They printed *Hansard* and *The London Gazette*, as well as order papers, bills and Acts, and other printed matter for Parliament.

Business Stationery

Lamson Paragon Ltd began in 1886 when the Paragon Cheque Book Co. was founded to make the cheque books originated by J.R. Carter in Canada. They were located in Fords Park Road, Canning Town, and built their factory there in 1893.

Newspapers

The first mention of the word 'newspaper' was in a German pamphlet entitled *Neue Zeitung vom Orient* celebrating victory over the Turks in 1502. In 1513, *Trewe*

Encountre told of the defeat of the Scots at Flodden Field and in England through-out the sixteenth and seventeenth centuries increasing numbers of pamphlets were published, albeit irregularly. Not that they received royal approval, for both Elizabeth I and James I discouraged their publication, James I decreeing the pro-hibition of 'all lavish and licentious talking in matters of state'.

And it was not only monarchs who took a dim view of the proliferation of pamphlets. Dramatists and playwrights – Ben Jonson included – complained of 'all the cheats of the lying stationers' and 'Captain Hungary who will write you a battle in any part of Europe at an hour's notice and yet never set foot outside a tavern'. More pointedly, those that objected to pamphlets observed that they 'wanted more harmless paper than ever did laxative physic'.

Members of Parliament were similarly reluctant for the public to read about their debates. *The Gentleman's Magazine* – founded by Edward Cave in 1731 – cleverly circumnavigated this by reporting in a sort of code language. It spoke of Parliament as 'The Senate of Great Lilliput' and one of its contributors, Samuel Johnson, referred to the Lords as 'Hurgoes' and the Commons as 'Clinabs'.

The first daily paper was *The Daily Courant*, printed at Fleet Bridge in 1702. It folded in 1735, after it had merged with *The Daily Gazetteer*. In 1712, stamp duty was levied on all newspapers and censorship was still rife, many editors finding themselves accused of seditious libel. The government would reward newspapers to gain their support. *The Times* was given £300 every year and also took bribes. But there were those who were brave enough to defy the government. For speaking out against the flogging of soldiers, William Cobbett was sentenced to two years in jail, as was Leigh Hunt in 1813 for criticising the Prince Regent in *The Examiner*.

The first Sunday newspaper was Johnson's *British Gazette and Sunday Monitor*, launched in 1779. The oldest surviving Sunday paper is *The Observer*, founded in 1791. *The Sunday Times* dates from 1822, and the *News of the World* (now defunct) from 1843. By 1785, eight newspapers were published every day in London, including the *Daily Universal Register*. It became *The Times* in 1788 and by the early nineteenth century, now popularly known as 'The Thunderer', was selling more copies than all the other newspapers put together.

As we have seen, stamp duty was abolished in 1855, and the price of newspapers fell from 6d to as low as ½d. Quick to seize the opportunity, *The Daily Telegraph* was launched soon after the repeal of the Act. By the 1870s, it had a circulation of 250,000, far more than *The Times*. Increases in literacy and the coming of the railways boosted newspaper sales enormously. W.H. Smith – based at the Fleet Street end of the Strand – supplied a distribution network. Their carts collected the papers, which were then packed and dispatched to waiting trains.

Early in the twentieth century, newspapers began to include photographs, and they very soon caught the public's imagination. An *Illustrated London News* cor-respondent commented: 'very early in my journalistic career I discovered the greater attraction of the picture as against the best possible writing matter. A royal wedding or a funeral added thousands to our sales, whereas the most exciting story or the most eloquent article left our subscribers cold.'

Newspaper and publishing magnates were enormously powerful and influential. The first newspaper baron was Alfred Harmsworth, later Lord Northcliffe. From humble beginnings, he began by launching small magazines: *Youth*, *Wheel Life* and *Bicycling News*, and then, in 1888, *Answers to Correspondents on Every Subject under the Sun*. He bought the *Evening News* in 1894 and founded the *Daily Mail*, 'the penny newspaper for one halfpenny', in 1896. By 1900, the *Daily Mail* – said acidly by Prime Minister Lord Salisbury to be 'printed by office boys for office boys' – had a circulation of 1 million compared to 38,000 for *The Times*. Carmelite House was built by Northcliffe to print the *Daily Mail* and *Evening News* in Carmelite Street (just off Fleet Street). Then, in 1903, Northcliffe founded the *Daily Mirror*, aimed specifically at women, and soon afterwards bought *The Observer*, *The Times* and *The Sunday Times*. Such was Northcliffe's prestige that he was offered a position in Lloyd George's War Cabinet. He turned the offer down, preferring instead the role of Director of Propaganda. After the war, Lloyd George was anxious for Northcliffe's support in the coming General Election. Northcliffe replied with typical bluntness: 'I do not propose to use my newspapers and personal influence … unless I know definitely and in writing and can consciously approve the personal constitution of the Government.' Such megalomania was too much for Lloyd George who – from the floor of the House of Commons and consequently with the protection of Parliamentary privilege – denounced Northcliffe's 'diseased vanity'.

Max Aitken, later 1st Baron Beaverbrook, was born in Canada in 1879. He made money there in various ventures and in 1910 set sail for England. He immediately moved into newspapers and acquired the *London Evening Standard* and then gained a major shareholding in the *Daily Express*. In the First World War, he was appointed Minister of Information, responsible for propaganda in Allied and neutral countries, in contrast to his fellow press baron Lord Northcliffe, who was responsible for propaganda in enemy countries. Always associated with the *Daily Express*, after the Second World War he turned it into the largest selling newspaper in the world with a circulation of 3.7 million.

The death knell of Fleet Street was sounded when Rupert Murdoch (News International) decided to move all his operations to newly built and equipped premises in Wapping and adopt new computerised technology. So began a protracted industrial dispute, starting on 24 January 1986 and lasting for over twelve months. After lengthy negotiations, 6,000 employees of *The Times*, *The Sunday Times*, *The Sun* and the *News of the World* were sacked, but despite persistent picketing outside what had become known as 'Fortress Wapping', the papers still came out. By the end of the 1980s, all other national newspapers had followed suit and left Fleet Street, chasing the lure of new technology.

Recently relocated to Luton, West Ferry Printers, in many ways a successor to Fleet Street, is one of the largest newspaper printers in Europe, employing 550 people. Each week, a staggering 21 million newspapers are printed including seven dailies, ranging from the *Financial Times*, *The Daily Telegraph* and the *Daily Express* to the *Daily Star* and *The Sport*. Every year, 211,000 tons of newsprint are used, imported from all over the world and stored in a special reel store in 1-ton rolls. The rolls are then conveyed to the presses (there are eighteen of them) by a fleet of robots that use radio signals to guide them to their next task. Pages are transmitted from the editorial offices and output directly to aluminium plates. Every night a fleet of 250 lorries leaves with the next morning's papers on board, ready to be distributed to wholesalers and then the retail trade.

Bank Notes

The Bank of England was founded in 1694 by the Scotsman William Paterson. Its first notes were handwritten but within a year William Staresmore, the Bank's stationer, began printing notes. At first, 1,200 were produced but they were soon withdrawn when a forged £100 note was discovered, prompting the instruction: 'all the Running Cash Notes for the future be written by some of the cashiers and that noe more of the printed ones bee delivered out'. In an attempt to get over the problem, watermarks were introduced.

Printing was strenuous work and the pressmen had to be strong in the arm. Damp paper was loaded onto a copper plate, engraved with the desired pattern, and had to be ink-free except in the engraved lines. The plate, with heavy pressure exerted, was rolled between two cylinders to give a note that had dark lines where the engraving was deep and lighter ones where it was shallow. The engraver was John Sturt, better known for his work with the Book of Common Prayer. It was a lengthy process and only four notes were made every minute. Nevertheless, this technique continued for well over 100 years. Forgery remained a problem. The Bank went to the length of dispatching a cashier to nearby Newgate Prison to consult with a forger, awaiting his date with the hangman. He recommended that the paper be treated with certain chemicals and as a reward for his co-operation he escaped the noose and instead was transported for life.

The Bank's first paper supplier was Alexander Merrial, who directed a paper mill at Sutton Courtenay, then in Berkshire, but after the forger's advice the Bank changed to a supplier of harder and better quality paper. This was Henry Portal, a Huguenot who owned a 'Company of White Paper Makers' at Bere Mill on the River Test near Whitchurch in Hampshire.

By 1731, James Cole was both engraving and printing bank notes at his premises at the sign of 'The Crown' in Great Kirby Street, Hatton Garden. At this time, the procedure for printing bank notes involved the moulds for paper making (made by Mrs Mary Smith, with forty years' service to her credit) being sent to Portal's premises. A representative of the Bank, 'the Bank's Officer', stationed at the paper mill and living in a house provided for him by the Bank, oversaw proceedings. When ready, the paper was stored in iron-bound chests and dispatched in Portal's wagons to Newbury and thence by the Kennet and Thames to London. The paper was stored at the Bank and each month an appropriate amount delivered to Cole. Then every morning the engraved copper plate was taken from

Printing bank notes at the Bank of England. *(By permission of Warren Grynberg)*

the Bank's custody to Cole's premises in Great Kirby Street. After printing, the cashier would count the sheets as they hung out to dry and when ten reams were ready they were taken to the Bank.

It was no surprise that the Bank's authorities were less than happy about work being done outside their walls and accordingly Cole was invited to relocate his business to within the Bank. This he did, but he still operated as a private concern and soon had eighteen presses in continuous operation.

Garnet Terry inherited Cole's business but the Bank remained concerned that they were dependent on a private concern. Therefore, in 1808, they concluded 'that the Business of Printing Bank Notes be taken into the Hands of the Bank and to be done within the Walls of the Bank by a Printer of their own'. Terry was duly sworn in and because he had given up his business was rewarded with a salary of £1,000 per year. The Bank, however, were quick to point out that 'the salary [was] not to be drawn into a precedent for any future printer'.

Forged bank notes continued to be a source of grave concern. Despite public outrage, many were executed for forgery; even being in possession of a forged note led to transportation. Many eminent men were consulted and presented many solutions, all more or less ineffective. They included Anthony Bessemer (brother of Sir Henry Bessemer of steel-making fame), Jeremy Bentham and Sir William Congreve, better known for his expertise on solid-fuelled rockets. A committee was eventually set up in 1817 to examine plans for the improvement of bank notes and for the prevention of forgery.

In 1832, the deputy governor of the Bank of England found himself in Ireland where he was introduced to John Oldham, chief engineer to the Bank of Ireland. Oldham's presses were steam-powered, obviating the need for strong pressmen; they also used a novel form of damping device for the paper. The Bank tried them out, were impressed and soon, in 1836, Oldham moved to London as 'Mechanical Engineer and Principal of the Engravers, Plate Printing, Numbering and Dating Office' and brought his presses with him.

With the discovery of the science of electrodeposition, Dr Alfred Smee introduced the technique of printing bank notes from electrotypes. An exact reproduction of an original mould – taken from the engraved plate – was produced and metal deposited in either recess or relief upon its surface. There were further advances in presses. With the Napier platen press, two boys were positioned at each end, the first to place the paper in the press and the other to remove the printed notes. The Bank also acquired the patent rights of Brewer & Smith's watermark, with its vignette designed by Daniel Maclise, well known for his frescoes in the Royal Gallery of the Palace of Westminster. The Napier press was eventually superseded by ones introduced by Henry McPherson in 1881. They remained in use until 1945. By this time the printing works had left its home in the precincts of the Bank and moved to St Luke's Printing Works.

The Bank had bought St Luke's Hospital – a former mental institution designed by F.W. Troup – in Old Street in 1917, and by 1920 reconstruction was complete. In 1956, there was a further move to Debden at Loughton in Essex. Today, De La Rue Currency produce bank notes at Loughton, where images are either engraved by hand onto metal plates or produced by computer-aided design (CAD). As many as eighty-five special inks are used and images duplicated onto the printed plates ready for the presses in Clerkenwell.

The life of a bank note is quite short, ranging from one year for a £5 note to five years for a £50 note. Used ones are shredded and taken for landfill.

17 Pottery

Pottery was known to ancient man, who at a very early date was aware of the three essential processes in its manufacture. They are mixing and moistening clay, forming the vessel, and firing the finished article to make it hard. The final temperature of firing, the type of clay and any ingredients added to it determine the nature of the finished product.

Broadly speaking there are three types of pottery: earthenware, stoneware and porcelain. Earthenware is clay fired at a low temperature (about 750–1,100°C). It is porous and has to be glazed to make it impervious to liquids. Stoneware is hard, impervious and made from a clay and flint mixture fired at a high temperature (1,200–1,400°C). Vitrification (the formation of a glassy outer surface) occurs at these high temperatures, so it does not have to be glazed unless a fine or coloured surface is required. Porcelain is the finest ceramic and originated in China in about AD 700. It is made from a fine, white clay, fired at a high temperature, and is translucent. In 1609, the Dutch East India Company imported Chinese porcelain to Europe, in particular to Delft, and it was there that potters imitated it with soft clay to make what has become known as Delftware. It was much in demand, and the Dutch exported it all over the western world.

Desmond Eyles, in his book *Royal Doulton 1815–1965*, tells us that Jacob Janson – who changed his name to Johnson when he came to England in 1571 from the Low Countries – built himself a pottery in Aldgate, where he made Delft pottery from tin enamelled earthenware. In the early seventeenth century, Christian Wilhelm is recorded as having a pottery at Pickleherring Quay (off Tooley Street in Bermondsey), and other potteries are mentioned in Southwark.

By the end of the eighteenth century, potteries and their associated kilns were found all over London: at Isleworth, Horseferry Road, the Hermitage at Wapping, Limehouse, Bankside, Nine Elms, Deptford, Greenwich, Mortlake and Bermondsey.

Fulham Pottery

On 23 April 1671, John Dwight obtained a patent, granted to him by Charles I, in which he claimed 'to have discovered the mistery of transparent earthenware commonly knowne by the name of porcelain or China and Persian ware, and also the misterie of the stone ware vulgarly called Cologne ware'. Dwight graduated at Christchurch, Oxford, and then spent time as secretary to three bishops of Chester before he founded the celebrated Fulham Pottery. He lived in Beare Street (now Fulham High Street) and is recorded, in 1673, as the 'master potter at Fulham'.

The Lord of the Manor granted him land to build his pottery, situated on some waste land between two inns, the 'Nagg's Head' and the 'Golden Lion'. Frequently known as 'the most famous name in English pottery', Dwight certainly impressed Robert Hooke, the renowned scientist, who noted in his diary: 'Saw Mr Dwight's English China, Dr Willis his head, A little boye with a hauke on his fist, Severall little Jarrs of severall colours all exceeding hard as a flint, very light, of very good shape. The performance very admirable and outdoing any European potters.'

The dealers in pottery were the Worshipful Company of Glass Sellers who sealed a contract with Dwight to buy exclusively from him and 'refuse the foreign'. The Fulham Pottery sold to the Glass Sellers 'all their fine brown bottles, jugs and other sorts of fine stoneware at £4 per hundred cast and coarse brown bottles, jugs and other brown stoneware at £3 per hundred cast'. They then promised 'not to encourage any persons to make or bring brown stoneware from beyond the seas'. For his part, Dwight promised 'constantly to keep the said Potthouse at work with able and sufficient workmen'.

The finest of John Dwight's work has been described as 'if not porcelain ... distinctly porcellanous'. But it seems that its manufacture was not financially rewarding. Dwight was in the habit of hiding away money, the tools of his trade and his recipes in secret, bricked-up compartments, in the hope that his descendants would be discouraged from attempting to make porcelain because of its poor financial reward. In 1673, Dwight's daughter, Lydia, died and soon afterwards he produced one of his earliest and best works: a figure of his daughter in hard stoneware with her head on a pillow, eyes closed and holding a bouquet of flowers. It is inscribed, 'Lydia Dwight, dyd March the 3rd, 1673' and is now housed in the Victoria and Albert Museum. Other fine works of Dwight are his life-size figure of Prince Rupert, busts of Charles II, Catherine of Braganza, James II and Mary of Modena, all to be found in the British Museum.

John Dwight died in 1703 and it is uncertain who continued running the Fulham Pottery, but in 1795 it is recorded that 'these manufactures are still carried on at Fulham by Mr White, a descendant in the female line of the first proprietor'. Ten years later John Doulton was apprenticed at Fulham and was destined to become one of the finest 'throwers' (one who shapes pottery by hand from a lump of clay thrown onto the centre of a revolving wheel) in the business. In 1812, aged 22, he got a job in a tiny pot-house in Vauxhall Walk, Lambeth.

Royal Doulton

The pot-house where John Doulton worked — one of twenty or so in Lambeth — was a small terraced house with a workshed and kiln in the back yard. Thirty years before it was run by a Mr Griffith and a Mr Morgan. They advertised in 1776 in Felix Farley's *Bristol Journal*: 'Wanted at Griffith's and Morgan's pot house, Lambeth, near London, a stone kiln burner, also a top ware turner and ingenious painter. These men must understand their business well, as the company have indifferent hands enough already.' Whether Griffith and Morgan were successful in their recruiting campaign is not recorded but when John Doulton joined the firm — now run by a Mrs Jones — they obtained one of the best throwers in London; within three years he was a partner, together with the foreman, John Watts. Mrs Jones had wanted to take her son as a partner but because he had 'clashed with the law', he fled to South America. (He was to return some years later and was given a job by Doulton as yard foreman, perhaps out of sympathy for the eccentric man's mother.)

In 1820, Mrs Jones retired and the firm became Doulton & Watts, but not before Watts had tried to call it Watts & Doulton. The matter was settled when the signboard outside read Doulton & Watts on one side and Watts & Doulton on the other. In time, however, Doulton came first. The firm specialised in making stoneware water pipes, glazed by throwing salt into the kiln when firing. They also made bottles, jugs, domestic utensils and Toby jugs – small brown jugs decorated with reliefs of characters, dogs, cottages, windmills and the like, made separately in moulds.

The business expanded over the years and eventually stretched from a point opposite the gateway of Lambeth Palace halfway to Vauxhall Bridge. It had many famous customers including Whitbread, the brewers; Berger's Paints; Crosse & Blackwell; and Warren's of the Strand, a boot blacking manufacturer. Charles Dickens, in his younger days, worked for Warren's and later called to mind pasting labels on thousands of Doulton & Watts blacking bottles.

The Great Reform Act came onto the statute book in 1832. The event was celebrated by Doulton & Watts making thousands of spirit bottles, known as 'Reform Bottles'. They depicted the head of William IV as well as the statesmen who had guided the bill through Parliament: John Russell and Henry Brougham. Each bottle was inscribed with slogans such as 'The True Spirit of Reform' and 'The People's Rights'.

In 1835, John Doulton's son, Henry, entered the business. He had been educated at University College, was well read (Pugin and Ruskin) and had a fine business mind. Ten years later he set up a separate factory in Lambeth High Street and began making glazed stoneware pipes for sewers and house drains. He soon expanded – helped in no small manner by the Public Health Act of 1848 – to make lavatory and wash basins.

The story of the manufacture of decorated pottery – Doulton Wares – begins when John Sparkes, an art teacher at Lambeth School of Art, approached Henry Doulton about the possibility of establishing an art studio at the Lambeth Pottery. The idea was firmly rejected by Doulton, prompting Sparkes to respond, with some bitterness, 'Let him stick to his drains.' In 1862, however, Edward Cresy, an architect, who owned a magnificent Rhenish salt cellar typical of those made in Cologne in the sixteenth century, showed it to Doulton as an example of what could be produced. Doulton was convinced, and decorated pottery began to be made at Lambeth.

Lambeth School of Art was awarded the first contract to make a number of terracotta heads of famous potters, such as Josiah Wedgwood and Bernard Palissy, to decorate the front of the factory building. They were designed by John Sparkes, with George Tinworth doing the work. Tinworth soon came to work at the Lambeth Pottery. He was born in Walworth in 1843, the son of a wheelwright. After attending Lambeth School of Art, he studied at the Royal Academy and remained at Royal Doulton until his death in 1913. Tinworth produced much fine work, including a terracotta reredos for York Minster. He was admired by John Ruskin, who wrote, '[his] work was full of fire and zealous faculty for breaking its way through all conventionalism to such truth as it can perceive'. In 1873, Doulton was employing six artists at Lambeth. By 1890, in what amounted to a rebirth of

The site of Brick Lane Brewery today.

Whitbread's Chiswell Street Brewery, 1792, by George Garrard.

Nicholson's tidal mills at Bromley-by-Bow.

Whitechapel Bell Foundry.

Turbine hall, Bankside Power Station, in the 1960s.

Bankside Power Station at night, in the 1970s.

Fire at Albion Mills. *(By permission of Southwark Local History Library)*

Former Billingsate Market, Thames Street.

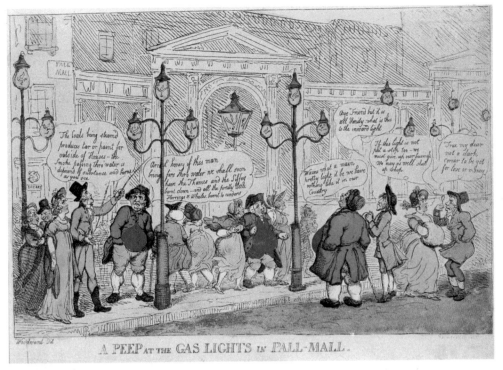

A PEEP AT THE GAS LIGHTS IN PALL-MALL.

'A Peep at the Gas Lights in Pall Mall.'
Frederick Winsor's gas-light display for
the Prince Regent, 1807.

Pottery kiln, Fulham.

The Coade stone lion, Westminster Bridge, formerly outside the Red Lion Brewery, Lambeth.

London Published May, 1834, by ACKERMANN & Cᵒ 96 Strand

Geometrical Elevation of

THE LONDON AND GREENWICH RAILWAY

Incorporated by Act of Parliament, 1833

as Designed & Projected by Geo. Landmann, Esqʳᵉ Engineer to the Company, & now Constructing

(This view represents that portion which crosses Corbett's Lane)

The London & Greenwich Railway. *(By permission of Southwark Local History Library)*

London Bridge Station in the nineteenth century. *(By permission of Southwark Local History Library)*

Shirley Windmill.

Carreras, Camden Town.

THE MINT.

Royal Mint. *(By permission of Tower Hamlets Local History Library)*

Royal Doulton
premises, Lambeth.
(Designed by
R. Stark Wilkinson)

Relief sculpture of potters by George Tinworth, Royal Doulton, Lambeth.

salt-glazed stoneware, the number had increased to 345, necessitating new premises to be built to house them: Tower Buildings, in Lambeth High Street.

Doulton showed a keen interest in the chemistry of salt glazing. Both he and Wilton Rix (the factory manager) met with Professor Arthur Church to study the chemical reactions that occurred during salt glazing and the influence on the process of the presence of sulphur and iron. During each new firing of a different formulation, Doulton looked on in wonder, 'like a farmer anxiously attending the birth of a calf', eagerly waiting to see the finished article.

The company won many awards, including the Grand Prix at the Paris Exhibition of 1878, followed by an award to Henry Doulton of Chevalier of the Legion of Honour. Queen Victoria was similarly impressed and as well as buying Doulton wares for herself, wrote to her daughter, the Empress of Germany, about them. The Court journal recorded that 'these productions have been considered worthy of special inspection by Her Majesty'.

The industrial ceramic side of the business expanded in parallel with the artistic. Pipes were exported all over the world and by 1879, Doulton had factories in Rowley Regis and Smethwick in the Black Country and at St Helens in Lancashire, producing 30 miles of pipes every week. The company received a further boost with the coming of the telephone by providing insulators for this new technology. Patent insulators were also supplied to the London Underground and such was the demand that Doultons were forced to grant licences to their competitors.

By the 1920s, Doultons was the only pottery remaining in Lambeth and, apart from Fulham, there were few in London. The company had other factories at

Erith, Paisley, Tamworth and Burslem, and more work was progressively transferred to these sites. The pottery in Lambeth closed in 1956.

Bow Porcelain

Porcelain was made in Bow, at a works called New Canton, which stood at the east side of Bow Bridge (crossing the River Lea) and north of Stratford Causeway. The factory was opened by three businessmen – John Weatherby, John Crowther and Alderman George Arnold. Those concerned with the hands-on manufacture of porcelain were Edward Heylyn and Thomas Fyre. They set up to 'manufacture porcelain in imitation of that imported from China'. A patent was taken out in 1744 for 'a certain material whereby a ware might be made of the same nature or kind and equal to, if not exceeding in goodness and beauty china or porcelain ware imported from abroad'. Bow Pottery's patents were far from precise. It is not clear what the 'certain' material was but it was said to be brought to Britain by a 'traveller' from the Cherokee nation of America. A later patent recorded the material to be 'virgin earth' produced by 'calcining animals, vegetables and fossils but some in greater quantity than others'. In 1758, the pottery was employing 300 people, 90 of whom were painters. And they recruited far and wide. Aris' *Birmingham Gazette* gave notice:

> … to all painters in the blue or white potting way and enamellers on china ware, that by applying at the counting-house near Bow, they may meet with employment and proper encouragement according to their merit; likewise painters brought up in the snuff box way, japanning, fan painting etc, may have an opportunity of trial, wherein if they succeed, they shall have due encouragement. N.B. At the same house a person is wanted who can model small figures neatly.

In contrast to porcelain made in Chelsea, Bow concentrated on 'the more ordinary sort of ware for common uses'. The Bow Pottery had retail premises in St James's and Cornhill. One of their customers was Benjamin Franklin who sent his wife, back home in America, 'a bowl remarkable for its neatness of figures made at Bow, near this City'.

Weatherby died in 1762, and one year later Crowther went bankrupt. Bow Porcelain Works went into decline and was finally bought by William Duesbury of Derby in 1775, merging with Chelsea to form the Crown Derby Porcelain Works, all manufacturing being carried out in Derby.

Chelsea Pottery

Fine porcelain was manufactured at Chelsea on a site at the corner of Justice Walk and between Old Church Street and Lawrence Street. How porcelain came to

Chelsea is uncertain. But the Victoria County History records that potters from Staffordshire were soon to arrive:

Carlos Simpson was born in Chelsea to which place his father went in 1747 (with others) ... they soon ascertained that they were the principal workmen, on whose exertions all the excellence of the porcelain must depend, they then resolved to commence business of their own at Chelsea and were in some degree successful, but at length owing to a disagreement among themselves, they abandoned it and returned to Burslem.

Others say it was first owned by the Duke of Cumberland and Sir Everard Fawkener, with Nicholas Sprimont, a former silversmith, as manager. Sprimont later became owner, and it is recorded, 'sometime in or about the year 1750 did build, erect, stablish several houses, warehouses, kilns, and other erections and buildings for the purpose of making porcelain and other wares at Chelsea'. Two years later, he took on Francis Thomas as his bookkeeper at a wage of £100 per year. Sprimont retired from active participation in the pottery in 1765, and was alarmed to find the profits of the firm plummeted thereafter. It would seem that bookkeeper Thomas – who died in 1769 – left a fortune of upwards of £7,000; far too much for a man earning only £100 per year. One can only put two and two together!

Chelsea Pottery's heyday was between 1750 and 1765 when they supplied the highest in the land. *The Daily Advertiser* of March 1763 noted: 'We hear that the elegant set of china, which Her Majesty ordered to be made at Chelsea and which is intended as a present for her sister, cost £1,150.' And, when Mrs Delaney visited the Duke of Cumberland's Lodge at Windsor, she witnessed a closet full of china and 'in the middle hangs a lustre of Chelsea china that cost six hundred pounds and is really beautiful'.

It is recorded – on the evidence of the factory foreman, who later found himself in St Luke's Workhouse – that none other than Dr Samuel Johnson had ideas that he could make improvements in the manufacture of china. But it was to no avail, for:

He was accordingly accustomed to go down with his housekeeper about twice a week, and staid the whole day, she carrying a basket of provisions with her. The doctor ... had free access to the oven and superintended the whole process, but completely failed, both as to composition and baking, for his materials always yielded to the intensity of the heat, while those of the company came out of the furnace perfect and complete.

In 1769, Chelsea was sold to the same William Duesbury of Derby who had bought Bow and, in 1784, he transferred all work to Derby. The two concerns were merged to form the Crown Derby Porcelain Works. Duesbury had been born in London in 1725 and practised as an enameller. He eventually became the largest manufacturer of porcelain in England.

Coade Stone

Particularly well known to this day in London is the ceramic work of Dorset-born Mrs Eleanor Coade. She used a particularly durable type of clay, resembling terracotta, known to her as Lithodipyra and to us as Coade stone. She had showrooms in Westminster Bridge Road and a factory where County Hall and the Festival Hall now stand.

The first stone works in the area was founded by Richard Holt who, in 1722, patented a process for manufacturing artificial stone and marble. Then, in 1770, Daniel Pincot styled himself an 'Artificial Stone Manufacturer' at his works by the King's Arms Stairs, Narrow Walk, Lambeth. It was Pincot who sold his works to Mrs Coade. The raw material was finely ground, and one of the factory's granite grindstones can still be seen to this day outside the Festival Hall. Once ground into fine particles, the material was cast into moulds and fired in a muffle furnace. Eleanor Coade died at the grand old age of 88 in 1796.

A feature of Coade stone was its ease of moulding and its resistance to weathering and erosion. Examples can be found all over London, in particular the lion that stands at the eastern end of Westminster Bridge. Originally painted red, it was made for the Red Lion Brewery which once stood at the end of Waterloo Bridge before it was demolished to make way for the Festival of Britain site.

Other examples of Coade stone in and around London are Nelson's pediment at the former Royal Naval College in Greenwich; Schomberg House in Pall Mall; Captain Bligh's tomb in St Mary's churchyard, Lambeth; and the rood screen at St George's Chapel, Windsor.

There is an urban myth doing the rounds that the composition of Coade stone is unknown. However, the online encyclopaedia Wikipedia tells us it is 10 per cent grog (finely crushed fired items), 5–10 per cent crushed flint, 5–10 per cent fine quartz, 10 per cent crushed soda-lime glass and 60–70 per cent ball clay from Dorset or Devon.

Coade stone reliefs from the former Danish-Norwegian Consulate, 1796.

18 Shipbuilding

It is perhaps difficult for us to appreciate today that until the late nineteenth century a flourishing shipbuilding industry thrived on the banks of the River Thames. In the medieval era it was supervised by the Worshipful Company of Shipwrights, which regulated work practices and imposed strict standards of workmanship. Edward III, in his assault on Calais in the Hundred Years War, had twenty-five ships built on the Thames, a number exceeded only in Dartmouth, where thirty-two were built. The shipbuilding yards thrived in the nineteenth century but with the advent of iron ships, the costs of bringing both coal and iron to London rendered them too costly and by the end of the century most firms had moved to the north or closed down. Those remaining concentrated on ship repair.

The Royal Dockyards at Woolwich & Deptford

There were other naval shipbuilding centres, of course – Sandwich, Winchelsea, Margate, Portsmouth, Dover, the Isle of Wight, Weymouth, Exmouth – but an expanding naval programme in the early sixteenth century persuaded Henry VIII to concentrate shipbuilding on the Thames. He was frequently resident at the Royal Palace at Greenwich, which undoubtedly influenced him to establish Royal Dockyards at nearby Woolwich and Deptford.

Woolwich was founded in 1513; Deptford four years later. One of the first ships to be launched at Woolwich was the mighty *Henry Grace à Dieu*, at the time the largest and most powerful warship in the world. She carried guns on two fully armed decks and was short rigged with four masts. Later, the Royal Dockyards, which also included Chatham, built many of the ships that defeated the Spanish Armada in 1588.

Many fine vessels were built in these two famous yards. In the early seventeenth century *Trades Increase* was launched from Woolwich. At 1,200–1,500 tons, she was the largest merchant ship ever built but, because she was not coppered, weeds tended to accumulate and clog up her bottom. On her first voyage she landed off the coast of Java (now part of Indonesia) so that the weeds could be removed; unfortunately, she came to the attention of the locals, who set her alight.

In 1612, Phineas Pett became the first Master of the Shipwrights' Company. Two years before, the celebrated naval architect designed HMS *Royal Prince*, a gift from James I to his son, Henry, Prince of Wales. She was 150ft in length, of 1,400 tons, carried 64 guns and was described as 'the greatest goodliest ship built in England'. The Pett family were architects to the Royal Navy for 200 years and in 1637 at Deptford, father and nephew, Sir Phineas and Peter Pett, built HMS *Sovereign of the Seas*, the 'finest warship afloat'. She was rated at 1,637 tons, with

Entrance to the Royal Dockyard, Deptford.

150 guns overall and five poop lanterns so large that ten men could stand upright in one alone.

Peter the Great of Russia came to Deptford to study shipbuilding in 1698. He stayed at Sayes Court, home of John Evelyn, where he apparently broke the place up after a heavy drinking session. His statue stands outside luxury flats at Deptford Creek.

As well as building warships, Deptford acted as a supply centre and built anchors, pumps and other equipment, while Woolwich had a much-praised ropeyard running along the site of present-day Beresford Street. Deptford was relatively close to the Navy Board Office in Seething Lane in the City and so was handily placed for experiments to be carried out on ship design, such as the coppering of timber ships to protect against clogging with weeds – the problem that had bedevilled *Trades Increase*.

By the end of the seventeenth century other dockyards had become more important and there were even plans to close Deptford. Philip MacDougall in his *Royal Dockyards of Woolwich and Deptford* gives their values in 1688: Chatham £44,940; Portsmouth £35,045; Deptford £15,760; Woolwich £9,669; and Sheerness £5,393. Samuel (later Sir Samuel) Bentham served his apprenticeship at Woolwich and, in 1800, rose to the post of Inspector General of Naval Workshops. He had a grandiose scheme to build a vast dockyard on the Isle of Grain and at the same time close Deptford, Woolwich, Chatham and Sheerness. It came to nothing but it highlighted a serious and growing problem that both

Deptford and Woolwich faced. Due to the build-up of silt, the channels at both yards were too shallow to take the larger modern vessels. Bentham delayed the inevitable by pioneering a river dredger which extracted 1,000 tons of silt every twelve hours, but it was really only a stay of execution. The last vessel built at Deptford was launched on 13 March 1869, and by the end of the month the yard had closed. HMS *Thalia* was Woolwich's last ship, launched on 13 July 1869, and a few months later that yard closed as well.

Blackwall Yard

In the late sixteenth century Roger Richardson leased from the Lord of the Manor, Lord Wentworth, 'a parcel of ground called Blackwall, extending from Poplar to the landing place there, in estimation 400 yards; also a parcel of waste ground from the said landing place to a sluice towards the south west in length 614 yards'. The lease later passed to Nicholas Andrews. It was here that Blackwall Yard was built.

Blackwall has a long history of both shipbuilding and repairing, and had particular associations with the East India Company. They unloaded their cargoes here onto smaller vessels which continued upstream to offload goods in the City at the Legal Quays. In the early seventeenth century, the Company's main shipyard was at Deptford, but Blackwall had a greater depth of water and accordingly William Burrell, the Company's shipwright, suggested that a dry dock, for ship building and repair, should be built there. It was known as Blackwall Yard. The first ship was the *Globe*, launched soon after 1612, the date when the yard began shipbuilding in earnest. Other ships were soon to follow: *Hector, Thomas, New Year's Gift, Merchant's Hope* and *Solomon*. Later in the seventeenth century, the yard passed to the Johnson family. Henry Johnson was a later member of that family, whose grandfather Peter Pett was, as we have seen, the navy's architect at the Royal Dockyard.

As well as building ships for the East India Company, Blackwall Yard in times of national need built warships for the Royal Navy. And in 1661, Samuel Pepys, English Admiralty Official, recorded in his diary: 'To a barge and went to Blackwall to view the dock which is newly made there and a brave new merchantman to be launched shortly called the "Royal Oak".' The association with the East India Company was never lost, however, and Johnson continued to make ships for them as well as for the navy. The same year that Pepys visited, Johnson commissioned George Sammon of Wapping to 'digg erect new build' a large wet dock of 1½ acres, the first wet dock on the Thames; not, it must be emphasised, for landing cargoes but for the fitting-out of vessels. Johnson and his Blackwall Yard thrived; he was knighted in 1679, and later passed the yard to his son who became MP for Aldeburgh. Pepys paid him a rather back-handed compliment, writing that he was 'an ingenious young gentleman, but above all personal labour, as being too well provided for to work too much'. The yard later passed to Captain John Kirby and then to the Perry family. Towards the end of the eighteenth century, John Perry II enlarged the dock to 8 acres and it became known as the Brunswick Dock, in honour of the Royal House of George III. Perry also built the famous Mast House, which allowed a large ship to have its masts lifted in under four hours.

As the years passed, there were others who had interests in Blackwall Yard, including John Wells from Deptford, George Green and Robert Wigram. Meanwhile, however, London's system of wet docks for cargo import and export began to be built – West India Dock in 1802, and London Dock in 1805. The East India Company wanted a dock for themselves so they purchased Brunswick Dock and, between 1803 and 1806, John Rennie and Ralph Walker constructed the East India Dock, two parallel docks for both export and import trade, with the export dock built on the former Brunswick Dock.

Shipbuilding continued at Blackwall and thrived: in 1840 there were 500 men employed, rising to 1,500 by 1866. But by now, ships were being built of iron rather than wood. Between 1612 and 1866, the Blackwall Yard built 357 wooden ships, 82 of them for the Royal Navy. Their first iron ship was *Superb*.

In 1843, the yard was split into separate firms: Wigram's were based in the western area, and Green's in the east. George Green's son Richard also built clippers, the *Challenger* being one of his most famous ships. It was a tea clipper built indeed as a challenge to American clippers, which at the time dominated the tea run from China. In 1850, Green exclaimed: 'I am tired of hearing about Yankee clippers and mean to build a ship to beat them.' His wish came true – *Challenger* beat *Yankee* by two days.

Wigram's built iron ships but closed in 1876; the site is now occupied by the ventilation shaft of the Blackwall Tunnel. Green's continued until 1907 and then, three years later, amalgamated with Silley Weir to form R. & H. Green and Silley Weir, ship repairers.

Thames Iron Works

The River Lea runs into the Thames at Bow Creek. It was to the western edge of the creek, in 1838, that Thomas Ditchburn, a shipwright, and Charles Mare, a naval architect, moved from their yard in Deptford. The site had been previously occupied by William and Benjamin Wallis.

The partnership of Ditchburn & Mare was short-lived; the pair argued and Ditchburn went his own way, leaving C.J. Mare in sole control. The other side of Bow Creek was marshy and inhospitable but Mare saw its potential and expanded his shipbuilding business to that – the Essex – side of the creek. It was a wise move because soon the North Woolwich Railway arrived, thereby enabling Mare to import raw materials by rail as well as water.

The firm, now known as C.J. Mare & Co., built ships of iron. But it did not stop making wooden ships. In 1849, it built the wooden paddle steamer *Vladimir* for the Russian government. Ironically, the vessel aided Russia in its war against the

British in the Crimea. For P&O, Mare built the *Himalaya*, at 3,438 tons the largest steamer of her day, and *Peru*, described as 'the most perfect steamer ever built on the Thames'. However, Mare was getting into financial difficulty and the yard, employing 3,000 men, was threatened with closure. The day was saved by Peter Rolt, a wealthy timber merchant and a descendant of Phineas Pett. The Thames Shipbuilding Co. Ltd was duly formed with capital of £100,000 in twenty shares of £5,000 each.

Berthed in Portsmouth Historic Dockyard, and still preserved for us to admire and enjoy today, is *Warrior*, the first iron-hulled battleship in the world and built at Bow Creek at the Thames Iron Works. The Crimean War had revealed the shortcomings of wooden ships. Furthermore, France was challenging Britain's long-held and jealously guarded naval domination. Prince Albert, after inspecting French capabilities at Cherbourg, stormed: 'The war preparations of the French are immense, ours are despicable. Our ministers use fine phrases but they do nothing. My blood boils within me.' It was against this background that *Warrior* was ordered.

The mighty ship, with four vast decks and a crew of 700, was built for speed and firepower. She was steam-powered but also had sails for day-to-day use. When under sail, her funnels would telescope down to avoid obstructing the sails. She was constructed at Bow Creek but her engine and boilers came from the firm of Penn's in Greenwich. On 29 December 1860, she was launched by the First Sea Lord, Sir John Pakington, and fitted out at nearby Victoria Dock. She was the largest, fastest and most powerful warship in the world – 380ft long, 58ft beam, 6,000 tons and 1,250hp – but never fired a shot in anger; no one dared challenge her!

Thames Iron Works was one of the most important shipbuilding yards in the country; so much so that it was said that 'the fleets of Europe were largely launched from the slips at Blackwall'. Vessels were supplied to Germany, Russia, Spain, Turkey, Portugal and Greece. The company's civil engineering department supplied iron for railway bridges and also the roofing for the Great Exhibition.

Thames Iron Works. *(By permission of Newham Local Studies Library)*

The building of HMS *Thunderer*, Thames Iron Works. *(By permission of Newham Local Studies Library)*

In common with all Thames-side shipyards, the company was beset with labour troubles and there was a series of strikes, culminating in a disastrous one in the late 1890s. The firm was later directed by Arnold Hill, who introduced a forty-eight-hour week, a profit-sharing scheme and a number of clubs and societies for workers to enjoy. There were science classes, a choral society, art and literature classes, a cycling club, amateur dramatics, a military band, a temperance league and – with lasting effect – a football club, founded and financed by Hill. In 1900, it became West Ham United Football Club.

The last warship built at the yard was HMS *Thunderer*, with its 13½in guns, at the time the largest dreadnought afloat.

In its later years, the Thames Iron Works supplied iron for Blackfriars and Hammersmith bridges before competition from the shipyards on the Tyne forced its closure.

J. Scott Russell & Co. and the *Great Eastern*

The owners of Thames-side shipyards often seemed to be playing some sort of musical chairs. After leaving C.J. Mare, Thomas Ditchburn opened a yard at Millwall (now known as the Isle of Dogs). It was taken over by the Millwall Iron Works and later Charles Mare joined it after he had relinquished his yard to the Thames Iron Works Co.

William Fairbairn, a Scottish-born engineer and shipbuilder, was a professional colleague of both Robert and George Stephenson and advised Robert on

the design of the Britannia Bridge across the Menai Strait. In 1835, he came to
Millwall and began to build iron ships, including many for the Admiralty. In the
adjacent yard another Scot, David Napier, was also building iron ships to innovative designs. Soon, yet another Scot arrived. He was Glasgow University graduate
and former Professor of Natural Sciences at Edinburgh University, John Scott
Russell. Acquiring the Millwall shipbuilding yard in 1848, he set about building, with Isambard Kingdom Brunel, the *Great Eastern* – the largest ship ever
launched on the Thames – for the Eastern Steam Navigation Co. The company
wanted to exploit the growing trade with Australia but was hampered by having
to maintain a series of coaling stations en route for the ships to refuel. The solution was the *Great Eastern*, which was designed to hold enough coal for the entire
round-trip as well as room for plenty of passengers and cargo. The ship's specification was impressive. She was 692ft in length, 83ft wide and had a displacement of
32,000 tons. She was able to carry 12,000 tons of coal at a speed of 14 knots with
800 first-class and 3,000 second-class passengers on board; alternatively there was
room for 10,000 troops. There were many Scots in the workforce of 2,000 who
built the ship at a cost of £920,000. The mighty ship was due to be released to
the river on 3 November 1857, with well over 100,000 people gathering both in
Millwall and across the Thames at Deptford to witness the sight. They came from
all sections of society – there were beggars, bankers, dignitaries of the church, old
women and young men – but all were to be disappointed! Brunel was the ship's
designer and he had devised a complicated system of hydraulic rams, chains and
brakes to launch his ship broadside, but she stubbornly refused to shift. One man
was killed in the process and others walked off the job. Charles Dickens was one
of the spectators and he wrote vividly:

> A general spirit of reckless daring seems to animate the majority of the visitors.
> They delight in insecure platforms; they crowd on small, frail housetops; they
> come up in little cockleboats, almost under the bows of the great ship. In the
> yard, they take up positions where the sudden snapping of a chain, or the flying
> out of a few heavy rivets, would be fraught with consequences that they have
> either not dreamed of, or have made up their minds to brave. Many on that
> dense floating mass on the river and the opposite shore would not be sorry to
> experience the excitement of a great disaster, even at the imminent risk of their
> own lives. Others trust with wonderful faith to the prudence and wisdom of
> the presiding engineer, although they know that the sudden unchecked falling
> over or rushing down of such a mass into the water would, in all probability,
> swamp every boat upon the river in its immediate neighbourhood, and wash
> away the people on the opposite shore.

After many more attempts, the ship was finally launched, sideways to the Thames, in
January 1858, to the sound of the bells of the church of St Nicholas, Deptford, ring-
ing in the background. It was all too much for poor Brunel. He was shortly to suffer
a fatal stroke. Regrettably, the *Great Eastern* was not a financial success. She worked

the Atlantic crossings for many years but is perhaps better known for laying the first transatlantic cable. Her end was certainly not as Brunel would have anticipated – she was used as a showboat with a floating circus and funfair at the 1886 Liverpool Industrial and Maritime Exhibition before being broken up in 1891.

Nelson Dockyard

Nelson Dockyard stood in Rotherhithe, on the south bank of the Thames at Cuckold's Point. The first occupier was John Deane, who built the 34-gun *Mary Galley* in 1687. Well known to William III, Deane returned to Russia with Tsar Peter the Great who visited England in 1698. There, he helped build ships in the Baltic.

Apart from a short period under Marmaduke Stalkartt, the yard was held by the Taylor, Brent and Randall families throughout the eighteenth century. It was John Randall, in 1789, who built *Discovery* for Commander George Vancouver – an exploration vessel that circumnavigated the globe in 1795 and whose captain discovered the island of Vancouver. Later, the ship was converted to a bomb vessel, equipped with a large mortar to fire explosive shells. The recoil was so great that the hull had to be particularly strong. It was therefore ideal for Arctic explorations where ice was an ever-present hazard. In another converted bomb vessel, the *Carcass*, Nelson explored the Arctic in 1772. *Discovery* ended her days off Deptford as a convict hulk!

John Randall's tenure at Cuckold's Point came to an unhappy end in 1802 with his suicide. There was a dispute with the Navy Board concerning a badly constructed ship, the firm's orders were down, and to compound his problems his workmen went on strike. Matters came to a head when one of his workmen hit him with a block of wood. All this was too much for poor Randall, who returned home and flung himself out of a window.

Thomas Bilbe was a later occupier of Nelson Dock. In 1854, he launched the *Orient*, to carry soldiers to fight in the Crimean War. Ships were hauled onto the slipway by a cradle, which originally was steam-powered until hydraulic power was installed in 1900. The shipbuilders of Rotherhithe never quite came to terms with the new ships made of iron. The year 1866 saw the last ship built at Nelson Dock, the *Argonaut*, after which the yard concentrated on repairing ships.

Thornycroft's

John Isaac Thornycroft was born in Rome in 1843 and came from a family of sculptors. The statue of Boudicca by Westminster Bridge is the work of Thomas Thornycroft, John's father.

In the early 1860s, John opened a shipbuilding yard at Church Wharf in Chiswick and, in 1862, constructed his first ship, *Nautilus*, a 36ft twin-cylinder-engine steam-powered launch with locomotive boilers. John, who had studied at Glasgow University and attended lectures at the School of Naval Architecture at Greenwich, applied scientific principles to shipbuilding. In 1869, he wrote a paper for the Royal Institution of Naval Architects which enabled the water resistance of ships to be calculated. Four years later, John Donaldson became a partner in the firm, enabling Thornycroft to devote more of his time to research.

Thornycroft built high-speed vessels. *Sir John Cotton*, ship number 18, built in 1874 for use on India's rivers and canals, had a speed of 21.4 knots and was described as 'the fastest vessel in the world'. He also built launches for torpedo attack. The first vessel ordered by the Admiralty in 1876 was HMS *Lightning*, and between 1874 and 1891 a total of 222 torpedo boats were built for the Admiralty and for colonial and foreign navies.

The company built *Peace* for the Baptist Missionary Society. Under the direction of George Grenfell, the celebrated missionary and explorer, she was dispatched in parts to Africa and reassembled there by Grenfell and local people. Her speed was 12mph, sufficient to outpace hostile canoes and, as Grenfell pointed out, fast enough for him not to end up in a cannibal's hot-pot. But *Peace* was captured by the Congo State Government in 1890 and it was only after a lengthy 'international incident' that the British flag was restored and the vessel handed back to Grenfell. *Peace* prompted the government to commission five other vessels in an attempt to relieve General Gordon in Khartoum, who was under attack from rebels. Much to the lasting displeasure of Queen Victoria – aimed particularly at William Gladstone – the orders were sanctioned too late and Gordon perished.

The devastating impact of torpedo boats caused the Admiralty to ask Thornycroft's to turn their attention to building vessels to destroy them – that is, ships fast enough to outpace them. *Daring* and *Decoy* were built and had a speed of 27 knots, well in advance of the Admiralty's specification of 21 knots. Other 27-knotters were *Ardent*, *Bruiser* and *Boxer*. In 1895, *Boxer*, at 29.31 knots, established a world record.

Rudyard Kipling was once on board the torpedo boat destroyer *Mallard* and recalled the occasion with his 1898 poem 'The Destroyers':

> The strength of twice three thousand horse
> That seeks the single goal;
> The line that holds the rending course,
> The hate that swings the whole;
> The stripped hulls, slinking through the gloom,
> A gaze and gone again;
> The Brides of Death that wait the groom,
> The Choosers of the Slain.

Thornycroft's became a public company in 1901 and John Thornycroft was rewarded with a knighthood, but by this time it was becoming increasingly obvious that the yard at Chiswick was unsuitable for the company's latest vessels. They had to pass beneath the Thames' bridges to reach the sea, which necessitated the upper parts to be left off and reassembled later. Accordingly, Thornycroft relocated to a yard on the River Itchen at Woolston, with its unimpeded access to the sea, and Church Wharf closed between 1903 and 1908.

Yarrow's

Alfred Fernandez Yarrow was born in 1842. At the age of 15 he was apprenticed to Messrs Ravenhill, manufacturers of marine engines for naval vessels. A talented youth, he added to his training at Ravenhill's by attending scientific lectures, including those given by Michael Faraday. His early life was far from easy: his father went bankrupt and life for the family was hard. When he completed his apprenticeship, Ravenhill's offered him a job at £100 per year, with the added comment: 'You had better accept it, for we know exactly the financial position of your father and you cannot do any better.'

But Yarrow wanted to work for himself and turned the offer down. He went into partnership with a Mr Hedley and set up as Yarrow & Hedley at Folly Wall on the Isle of Dogs. Their first order was for a thief-proof door for a safe in a jeweller's shop in Brighton, but eventually they began making steam launches; 350 were built between 1868 and 1875.

The firm soon outgrew Folly Yard and moved to London Yard, a short distance downstream in Cubitt Town. The site was first occupied by the chain cable

Torpedo boat.

works of Brown, Lenox & Co., and then by the boiler makers and ironworks of Westwood, Baillie, Campbell & Co. The yard had a 450ft river frontage, and new buildings were built with steel from Sir William Arrol & Co.

Yarrow's built launches for torpedo boats. Their first order was from the Argentine and other orders from foreign governments followed. They soon began supplying the Admiralty with a series of high-speed boats: the *Batoon*, the *Hornet* and the *Havock*. The *Hornet* attained a speed of 27.3 knots. It was powered with water tube boilers, a marked improvement on the previously used locomotive boilers. Then, by using high-tensile steel rather than mild steel, the firm was able to reduce the thickness of the ship's hull and so attain still greater speeds. A Yarrow vessel was the first to attain 30 knots.

In 1895, orders for eight destroyers and ten torpedo boats were received from Japan. In their attack on Port Arthur, the Japanese were thus able to sink the Russian battleships *Czarevitch* and *Retvisan*. Yarrow's thus played a decisive part in Japan's victory in the Russo-Japanese War of 1904.

Yarrow's moved to the Clyde in 1907. C. & E. Morton, manufacturers of pickles, jam and soup, replaced them in London Yard (see p. 229).

Other Shipyards & Marine Engines Manufacturers

Many shipbuilders made steam engines as well as complete vessels; others confined their activities to engines. C. Stansfeld-Hicks, in his series of articles in the Port of London Authority magazine *PLA Monthly*, detailed the many firms which manufactured on the Thames, and much of the following is taken from his researches.

John and Robert Batson inherited a yard, just downstream of Limekiln Dock, in 1751. It was previously owned by their uncle Robert Carter, described as a 'shipbuilder of the hamlet of Poplar'. In the eighteenth century, Batson's built seventeen ships for the Royal Navy, including *Captain* on which Horatio Nelson fought at the Battle of Cape St Vincent. Nelson later commanded *Janus*, another of Batson's ships, in 1780. She was described as a 'fine frigate, quite new'. In 1762, at the tender age of 20, George III's brother, Edward, Duke of York, captained *Phoenix*. Both of the Batson brothers were dead by the early nineteenth century and, in 1817, Johnstone's Directory says the yard at Limehouse Hole was in the hands of Cox, Curling & Co.

J. & A. Blyth were at Limehouse and were associated with the sugar planters in the West Indies. They eventually closed in 1876.

Messrs Seaward & Capel made engines at the Canal Iron Works, Millwall, for both river craft and ocean-going vessels, including for the Royal Navy. In about 1860, they became Jackson & Watkins. Taken over by the East & West India Dock Co. in 1890, they ceased operating thus allowing the entrance to the South West India Dock to be enlarged.

Lewis & Stockwell were at Blackwall Point, and Barnes & Miller began in Stepney at Glasshouse Fields. They later became Miller & Ravenhill, between

Forrest's lifeboat builders, Limehouse. Note the viaduct of the London & Blackwall Railway.

1840 and 1850, building river steamers, and then Miller, Ravenhill & Salkeld. The firm closed in the 1870s.

Charles Langley had his own yard but later moved to Millwall Iron Works as managing director.

J. & G. Rennie, descendants of the famous John Rennie, operated from Norman Road in Greenwich. First they made marine engines and later ships. They moved to Wivenhoe in 1911.

Between 1843 and 1885, Samuda Brothers made ships for foreign navies, and steam packets which ran from Holyhead to Ireland.

The Royal National Lifeboat Institution (RNLI) bought lifeboats from Harton of Limehouse and also Forrest's. Later, in 1895, the Thames Iron Works took over and became the official suppliers of the RNLI.

Henry Maudslay is best known as the inventor of the lathe, an innovation that revolutionised engineering. Maudslay, Sons & Field were founded in 1810 at Lambeth and built engines, including for Brunel's *Great Western*. They were known as the finest marine engineering firm in the world, and at one time 70 per cent of the engines built for the Admiralty came from their yards. They also had a shipyard at Greenwich and continued in operation until 1900.

Dudgeon's, established in 1855, and well known as pioneers of twin-screw ships, were first at the Sun Iron Works in Millwall and later moved to Cubitt Town.

Messrs Humphrys & Tennant at Stowage Wharf, Deptford, made engines for the navy. They began in 1852 when Edward Humphrys left John Penn & Son (see below).

The firm of T.A.Young was at Orchard Place, Blackwall, between 1857 and 1893.

Brent's, in Rotherhithe, built *Rising Star*, influenced in its design by the ubiquitous Richard Trevithick. It was the first steamship to cross the Atlantic in an east–west direction. William Elias Evans was also at Rotherhithe, and Pitcher's was at Blackwall.

John Penn & Son began at Greenwich, first making treadmills to work the cranes in the docks, then the windmills that surrounded the Isle of Dogs. They also made the finest engines. Many were supplied to the navy as well as to foreign governments. The firm was purchased by the Thames Iron Works in 1899.

In addition to those mentioned, there were scores of smaller yards building river boat steamers at Rotherhithe, Blackfriars, Westminster, Limehouse, Nine Elms, Battersea, Hammersmith and Lambeth.

19 Textiles

The Silk Industry

In 1685, under the French King Louis XIV, the Revocation of the Edict of Nantes caused difficulties for religious minorities. As a consequence, many thousands of French Protestants who wished to remain faithful to their beliefs sought refuge in Britain. Known as Huguenots, many were skilled silk weavers from Lyons and Tours (the centres of the industry), and brought their skills in the manufacture of lustrings, velvets, brocades and satins to Spitalfields and its surrounding area. And they were welcomed, as Stow confirms: 'Here they found quiet and security and settled themselves in several trades and occupations; weaving especially … a great advantage hath accrued to the whole nation by the rich manufactures of weaving silks, which art they brought along with them.'

Many arrived without a penny to their name. To provide relief, an Order in Council was passed in 1687, decreeing that a collection, known as the Royal Bounty, be set up. The Huguenots had their own committee to distribute the money (£16,000 per year). Their first report told of over 13,000 settling in the Spitalfields area.

There was a great demand for silk – previously it had to be imported from France – but, over time, fashions changed and other fabrics were introduced from abroad. The Huguenots began pressing for a ban on the import of foreign silks, often with little success. Matters came to a head in 1713, when a mob of Huguenot weavers threw ink over women wearing Indian calico or linen clothes. The Trained Bands were called out and protesters found themselves in the Marshalsea or Newgate prisons. In 1764, further trouble arose when a large group of weavers marched on Parliament 'with drums beating and banners flying', insisting on a complete ban on foreign imports. J.H. Clapham, in his article 'The Spitalfields Acts 1773–1824', tells us how 'bands of weavers traversed the streets at midnight, broke into the houses and destroyed the property of the manufacturers' and how 'an informer was dragged to a pond, thrown in and was pelted to death'.

The Huguenots did not forget their Protestant heritage. In 1745, when London faced the possibility of an invasion by the Catholic Bonnie Prince Charlie, 2,919 French weavers gave an assurance that they would 'take up arms when called thereto by His Majesty in defence of his person and government'.

James Leman and Anna Maria Garthwaite were well-known silk designers, and some of their work is housed in the Victoria and Albert Museum. Often silk designers and weavers would work for silk mercers. One such was a Mr Carr who supplied a lady living in St James's Palace. She wrote to her sister in 1750: 'In the morning I had Mr Carr and silks, and I bought a rich satin for a sash, as near the colour of the ribbon we bought together as I could get it and a purple and white flowered silk. Mr Carr was a silk merchant and Anna Maria Garthwaite designed for him.'

Towards the end of the eighteenth century, disputes arose between journeymen and master weavers over wages. Parliament therefore passed a series of statutes – the Spitalfields Acts of 1773, 1792 and 1811 – whereby Aldermen of the City and Magistrates of Middlesex were given the authority to fix wages. The Acts turned out to be somewhat of a mixed blessing; cheaper cotton began to replace silk, and because silk workers' wages were fixed at a level too high for the market to support, many looms and their operators lay idle. Sir Thomas Fowell Buxton (of Truman's Brewery fame – see Chapter 24) spoke up in sympathy for the weavers' distress: 'It partook of the nature of a pestilence which spreads its contagion around and devastates an entire district.' The Spitalfields Acts were eventually repealed in 1824.

The introduction of the Jacquard loom in the mid-nineteenth century brought relief to Spitalfields. Now figured silks could be produced. At the time, there were approximately 15,000 looms in the area and as much as half the population depended on the silk industry for their livelihood.

In 1837, in a report to the Poor Law Commissioners, Dr Kay tells of a typical weaver:

> A weaver generally has two looms, one for his wife and another for himself, and as his family increases the children are set to work at six or seven years of age to quill silk; at nine or ten years to pick silk; and at the age of twelve or thirteen (according to the size of the child) he is put to the loom to weave. A child very soon learns to weave a plain silk fabric.

But decline was to come, and from about 1860 onwards – when the duty on imported silks was abolished – the number of weavers decreased. The 1901 census revealed only 548 working in the Spitalfields area, but weavers' houses in Spitalfields are still there for us to admire today in Fournier Street, Wilkes Street and Princelet Street.

There were silk weavers in other areas of East London. Leny Smith had a factory at Hackney Wick in 1787. By 1811, he was employing over 600 women and was said to be the largest producer in the country. The firm also had premises in Taunton.

Calico Printing

Calico printing was brought to England by William Sherwin, who took out a fourteen-year patent in 1676. He traded from West Ham, which became a centre of the industry. Calico takes its name from the Indian city of Calcutta where the technique of dyeing cotton cloth with colours originated. The so-called 'Calico Grounds' were situated between Stratford and Abbey Mills and employed 360 hands in 1811. Thereafter the industry went into decline.

Sweated Labour in the Clothing Trade

The use of sweated labour was widespread in the clothing industry. It was familiar to Charles Dickens – Jenny Wren in *Our Mutual Friend* was a seamstress. And Henry Mayhew's writings bring to life the degradation and appalling conditions of so many of London's poor who were trapped by it. Sweated labour was centred in the Spitalfields area.

Sweated labour was synonymous with long hours, unhealthy surroundings and work often done by outworkers in unregulated premises. In the mid-nineteenth century, it became endemic in the tailoring trade. At the beginning of that century, however, tailors had been relatively well off. There were two sections of the trade: the 'flints' and the 'dungs'. The flints had regular employment in their master's workshop and were paid a fixed wage; the dungs, in contrast, worked at home, under piecework. The flints far outnumbered the dungs.

But manufacturers realised – because of the vast increase of cheap labour provided by women or East European Jews fleeing the Russian pogroms – that costs could be cut by employing outworkers. Mayhew reveals that of the 21,000 tailors in London in 1849, 3,000 were in the 'honourable' trade (flints), while the remainder laboured wretchedly in the slop trade in homes and hideous workshops, often working an eighteen-hour day. Conditions were dreadful: many tailors and their families worked and lived in one room. This was reflected in the death rates, which were double that of agricultural workers. Pay was typically half that of an 'honourable' tailor working in the West End.

In the 1860s, the sewing machine first appeared in Whitechapel and Covent Garden. But, rather than inducing manufacturers to move production to factories, outworkers were forced to obtain Singer machines on hire purchase or rent them. Singer had thirty or so 'rent collectors' in the East End who went from house to house to get weekly payment. Mechanisation brought other changes. Specialisms arose whereby workers made just one particular item; for example, sleeves, pockets or lapels.

Women came off worse. The *Women's Trade Union Review* wrote:

> Women were able in these days of machinery to learn in 14 days all that is required of them to manage their machines, and they not only hurt themselves

by the wages they accept, but injure their husbands and brothers by undertaking to do for 5/- what men can get 13/- a week for and they therefore push out of employment.

Clothing Firms

Simeon Simpson opened his first clothing concern in 1894. Later, in 1917, he opened a factory in Middlesex Street (better known today as Petticoat Lane). Then, in 1929, because of increased demand, he built a model factory at 92–100 Stoke Newington Road and, in 1934, introduced the popular DAKS range of clothing. Over 2,000 people were employed at Stoke Newington. Alec Simpson took over the running of the firm when Simeon died and opened the famous Piccadilly store, Simpson's. The model factory was bombed in the Second World War and, in 1948, the company built a new factory in Scotland.

Tapestry Works

It was becoming increasingly fashionable in the sixteenth century for the well-to-do to decorate the rooms of their homes with tapestries rather than wooden panelling. Both Henry VIII and Elizabeth I employed tapestry workers.

There was a tapestry workshop in the early seventeenth century – albeit short-lived – in Warwickshire, but the early Stuarts looked to the workshops of Paris for tapestries. The Stuarts were impressed by them and accordingly a commission was formed to investigate whether tapestry could be weaved in England. As a result, Sir Francis Crane was appointed to build a factory, which was granted a twenty-one-year monopoly and staffed with fifty Flemish weavers. In 1619, he settled on a site in Mortlake, where he purchased premises previously occupied by Dr John Dee, Elizabeth I's physician.

Many famous men were associated with the Mortlake factory. Crane was secretary to Prince Charles (the future Charles I) and MP for Penryn and Launceston. He was also patronised by Charles' favourite, George Villiers. Both Van Dyck and Rubens may have contributed to Mortlake designs. The first set of nine pieces to be woven in 1620 was 'Vulcan and Venus'. Then, 'certayne drawings of Raphael of Urbino, which were designed for tapestries, made for Pope Leo X and for which there is £300 to be payed, besides their charge for bringing home' came to Mortlake. Seven of Raphael's ten cartoons were reproduced as tapestries.

Crane died in 1636 and was succeeded by his brother, Richard, but unfortunately there were debts to be paid and the workmen went without wages. All this resulted in Charles I buying the Mortlake concern for £5,811 11s 6d and the works were renamed 'The King's Works'.

The Civil War began in the early 1640s. It was a disaster for the Flemish weavers at Mortlake. Once more they went without pay. They wrote: 'Also we cannot return to our birthplace, for such religion rules there that we dislike, and through the knowledge the Lord has given us we have joined thus our church here, and so if we returned, that would be like a dog returning to its own vomit.' And, one year later: 'We are entirely without work.'

After the Civil War, copies of Mantegna's 'Triumphs of Caesar' (now on display at Hampton Court) were made and tapestries woven. Also at Hampton Court, Oliver Cromwell's apartments were lavishly fitted out with Mortlake tapestries. But Mortlake never recovered its former glory. Charles II did not have the artistic fervour of his father, and although the factory continued for a while, it eventually folded in 1703.

Towards the end of the seventeenth century, Walloon refugees set up a tapestry works in Fulham. Also in Fulham, in 1753, the Frenchman Peter Parisot, patronised by the Duke of Cumberland, began weaving carpets and tapestries. The tapestries were made 'after the manner of the Gobelins', and the carpets 'after the manner of Chaillot'.

20 Other Establishments & Trades

The Royal Arsenal

The Royal Arsenal developed from the Royal Dockyard founded by Henry VIII at Woolwich. It was at Woolwich Dockyard that Henry's mighty ship, the *Henry Grace à Dieu*, was launched in 1515, and this ship, along with all others, had to be fitted out with cannon and munitions. This was done from the adjacent Gun Wharf which had a river frontage of 265ft, an ordnance storehouse and a rope yard.

The area became known as The Warren and contained a grand house called Tower Place. It was acquired by the government in 1671 from Sir William Prichard, a merchant tailor. In 1696, a laboratory opened at a cost of £2,961. The Brass Foundry was built in 1716, and later, in 1803, the Carriage Department. In time, they became the Royal Laboratory, the Royal Gun Factory and the Royal Carriage Department, the whole establishment becoming – by decree of George III – the Royal Arsenal. Many famous engineers worked at one time or another at the Arsenal, including Henry Maudslay and Marc Brunel, and it was Brunel who installed steam-driven saw mills. The Royal Carriage Department was responsible for the manufacture of gun carriages for both land and naval use. The noted chemist Sir Frederick Abel worked in the Royal Laboratory on gun-cotton and cordite explosives.

The tens of thousands of people who worked at the Royal Arsenal have left their mark in areas other than ordnance. In 1828, a small group of workers founded an association to buy food. It became the Royal Arsenal Co-Operative Society. There was also football. In 1886, another group of men formed a football club, playing their first match in December of that year against a team on the Isle of Dogs. They won 6-0 and were known as Woolwich Arsenal, moving to north London in 1913 to become the famous Arsenal Football Club, the Gunners.

At its height, during the First World War, the Royal Arsenal employed about 80,000 workers. After the war, steam locomotives were built, but well before the Second World War it was realised that further ordnance factories were needed. Nevertheless, 30,000 workers remained at Woolwich during the Second World War. Apart from increased activity during the Korean War, the Royal Arsenal began a process of contraction in the latter half of the twentieth century, and in 1994 ceased all operations.

Royal Small Arms Factory

Before the Royal Small Arms Factory was established, the government bought its guns from private suppliers. Many of them came from Birmingham, but by the beginning of the nineteenth century many skilled workers had emigrated from that city and the government was forced to buy from abroad. Lord Chatham summed up this unsatisfactory state of affairs thus: 'The knowledge of making fire arms has become almost extinct.'

In response, in 1804, the Board of Ordnance opened a small armoury in Lewisham to make gun barrels. A little later, Captain John By of the Royal Engineers was dispatched to survey land on the River Lea near to the Gunpowder Works at Waltham Abbey. He reported that on the river at Enfield, 'there was a fall of nearly eleven feet with a great supply of water, supposed to be equal to 80 horse power', making the site ideal for a factory. The Royal Small Arms Factory was completed in 1816 and had three mills with water wheels to power the site.

The first job of the factory was to make 'Brown Bess' muskets, a flintlock musket which, from the mid-eighteenth century to the 1830s, became the main musket of the British Army. Rifles and swords were also manufactured. Men were transferred from Lewisham and others recruited locally. They lived on site, were known as 'Lockies' (after Enfield Lock) and had every convenience provided for them, including a pub. The pub was originally a farm, the 'Swan & Pike'. It was reasoned that the tenant should be given a licence to run his establishment as a public house, because 'it would be more advisable for workers to have access to a house, the occupier of which was a tenant absolutely under the control of the Board, and subject to be displaced for bad conduct, than to have the men going off to other public houses a mile or two distant'.

The men were well paid and it is a matter of opinion whether the ploy worked or not, for an 'Old Inhabitant of Enfield' wrote later: 'High wages was to many of the skilled men a curse, instead of being a blessing. The men divided their week into two portions – three days work and four days riot.'

The Royal Small Arms Factory expanded significantly at the time of the Crimean War and by 1860 was making 1,744 rifles every week. Almost all guns had 'EN' – a reference to Enfield – in their name: the Lee-Enfield rifle, the Bren gun and the Sten gun. There was further expansion in the twentieth century, but by the 1960s many workshops had closed. The Royal Small Arms Factory became part of British Aerospace and eventually closed in 1988.

Gun Making

An early London gunsmith was John Twigg. He was born in Grantham in 1732, and learnt his trade from an Irish gunmaker. He moved to London and worked at various addresses (Angel Court, Charing Cross; the Strand and Piccadilly). He is often referred to as the 'father of the duelling pistol'. Durs Egg, born in Switzerland and father of Augustus Egg, the noted Victorian painter, worked with Twigg.

Most eminent of all in the later eighteenth century was Joseph Manton, who pioneered many advances in firearms. He was skilled in making barrels, made advances in shot design and improved duelling pistols. His workshop was in Oxford Street, but Manton's career was blighted by a long-running dispute with the Board of Ordnance. He designed a wooden cup which allowed quicker reloading and accuracy of cannon. The army were interested but fell out with him over payment. Manton wanted a lump sum of £30,000; the army offered one farthing royalty for every shell produced. It ended up with Manton going bankrupt.

James Purdey worked for Manton and in 1814 opened his own business at 4 Princes Street, Leicester Square. After Manton went bankrupt, Purdey moved to his premises in Oxford Street. Later in the nineteenth century the firm, now run by James Purdey the Younger, moved to its present premises in South Audley Street and Mount Street. Purdey's have many famous customers and obtained

Purdey's gun and rifle manufacturers, Mayfair.

their first royal warrant in 1868, when they supplied the Prince of Wales (the future Edward VII).

Boss & Co. trace their history to 1773, when William Boss was apprenticed to a Birmingham gunmaker. He moved to London and, like Purdey, worked for Joseph Manton. His son, Thomas, was soon to join him at Manton's and served his apprenticeship there, after which he started his own business at 73 St James's Street. The firm always prided themselves on making the best guns, but they were not cheap. George VI observed: 'A Boss Gun, a Boss Gun, bloody beautiful but too bloody expensive.' In 1908, Boss & Co. moved to Dover Street, then Albemarle Street, back to Dover Street, and are now in Mount Street.

Rope Making

Rope was an essential material in London's Docks. It was made from manila and hemp. Rope was made in Limehouse by Huddart & Co. Joseph Huddart (1741–1816) was a hydrographer who surveyed coastlines and harbours. He is best known for the development of steam-driven machinery for laying up and binding rope, thus automating what was traditionally a labour-intensive process. His rope was vastly improved in quality and reliability and set the standard for all future rope making. Ropemakers Fields, a park laid out by the London Docklands Development Corporation in Limehouse, is a reminder of the industry that until the early twentieth century thrived here. There are rope mouldings on the park's railings – another nice reminder of former days.

Brush Making

Brushes were often made by homeworkers but there were factories as well, such as G.B. Kent & Sons. Established in the eighteenth century, they had premises in Great Portland Street, but in the nineteenth century moved to Robinson Road and Approach Road near Victoria Park. All sorts of brushes were made and, in 1882, the firm employed 600 people, many of them making toothbrushes. The bone handles came from the legs of bullocks and 600 were typically slaughtered every week to make in excess of 8,000 toothbrushes. Other brush makers were S. Ludbrook & Co., whose Bancroft Brush Works were in Mile End opposite where Queen Mary College now stands; and Arthur Dellow & Co., makers of baskets as well as brushes in Commercial Road, Shadwell.

The Fur Trade

The fur trade does not have the best reputation. It was, however, firmly established in London. Seals migrate by swimming from the warm waters of the

Alaska factory, Bermondsey.

South Pacific to the Pribilof Islands off Alaska. There, men known as seal jackers worked from a parent ship in rowing boats and killed indiscriminately. In 1796, a Southwark trunk manufacturer, Thomas Chapman (a group of rocks in Antarctica are named after him), learnt how to remove the long top hair of seals and leave the whole of the original fur firm on the skin. He duly began processing fur seal skins for the hat industry. Other garments were made such as gentlemen's waist-coats and ladies' jackets.

In London, Dresser & Dyer began at 26 Poultry in 1823. Founded by John Moritz Oppenheim, they later moved to Basing Lane, off Bread Street, and were near neighbours of the well-known furriers P.R. Poland. Oppenheim processed many thousands of Alaskan furs at a warehouse in Cannon Street and a factory at Castle Yard, Blackfriars. Soon, Charles Walter Martin was put in charge and a move was made to the Alaska Factory at 61 Grange Road, Bermondsey. As with so many industries, workers had specific jobs – there were shavers, blubberers, flesh-ers, dyers and tubbers. It was the job of the tubbers to dress the skins by stamping on them with their bare feet in tubs. But by the end of the nineteenth century the fur industry was in decline and the tubbers – to preserve their jobs – were said to be 'stamping on nothing'. There is a story that, in response, Mr Vernon Martin, now in charge, chased them trouserless across the yard with a broom.

Martin's then diversified by processing Chinese skins and dyeing bearskins for the Guards. Their head office was in Thames Street where seven of the nine floors were used for displaying, sorting and grading an extensive range of skins. The Alaska Works also processed Russian hare, Tibetan lamb and Chinese goat. Martin's took over a works in Pages Walk, Bermondsey, and expanded to 239 Long Lane, Bermondsey. Just before the Second World War they were employing 1,100 workers.

Cigar & Cigarette Manufacture

Tobacco was imported into the London Dock and stored in the New Tobacco Warehouse, a Grade I listed building by Daniel Asher Alexander. It was built by Napoleonic prisoners of war in 1814, and still stands today for us to enjoy. The government was quick to appreciate the revenue-raising potential of tobacco and by 1826, £3.5 million had been raised in taxation. Huge amounts of tobacco were stored and despite the stringent security of high walls and locked doors, theft was common. All the imported tobacco was carefully inspected and any damaged cut away and burnt on site in a kiln, known locally as the 'Queen's tobacco pipe'. Capacity was vast and as much as 24,000 hogsheads of tobacco could be stored, Customs duties being paid before dispatch.

Nearby, in Commercial Street, was the cigar and cigarette manufacturer Godfrey Phillips, founded in 1846 and at one time employing over 2,000 women. Rosemary Taylor and Christopher Lloyd describe in their book *The East End at Work* how cigars were made at J. & S. Hill's in Shoreditch: a leaf of tobacco was spread out on a bench and gashes made in it. One of the workers (all sitting side by side) took a few pieces of tobacco and rolled them in a cigar shape before placing in an iron guide to cut for length. The cigar was completed by rolling it and one end twisted to secure it.

Just south of Mornington Crescent Underground Station and at the northwest end of Hampstead Road is the former Black Cat Factory of Carreras, cigarette manufacturers. Although the company claimed to date from 1788, their origins are far from clear. The family were of Spanish origin and settled in Somers Town, a refuge for many from Spain. Dickens, in *Bleak House*, refers to 'poor Spanish refugees, walking about in cloaks, smoking little paper cigars'.

José Joaquin Carreras is recorded as a 'cigar importer and snuff manufacturer' with a business in Wardour Street, Soho. He had eminent customers, including the Prince of Wales (later Edward VII) and the 3rd Earl of Craven. The earl would experiment with different sorts of tobacco from which the famous (and throat-scouring) brand Craven 'A' is derived.

By the 1880s, the business had been bought by the American financier William Johnston Yapp, later to be joined by Bernhard Baron, a Russian-born American. He had invented a cigarette-making machine capable of producing 18,000 cigarettes per hour – many more than the 300 or so that could be made by hand. To begin with, the company had a factory in Aldgate and then, in 1910, opened in City Road near to Dingley Road. Expansion forced a further move and, in 1916, they opened their Arcadia Works in Mornington Crescent. The factory was designed by M.E. & O.H. Collins in the Egyptian Revival style. Outside are two 10ft-high bronze cats – the firm's trademark. In its heyday 3,000 workers were employed, most of whom were women. In 1958, Carreras was acquired by the Rembrandt Tobacco Co. and soon moved to Basildon in Essex.

Omnibus & Coach Making

The firm of Adams & Co. made horse-drawn omnibuses at their Fairfield Road works in Bow. With the coming of the railways, the firm began making railway carriages. Hooper & Co. made horse-drawn carriages for Queen Victoria and the Prince of Wales at their Little Windmill Street manufactory. They won prizes at the Great Exhibition of 1851 for their 'elegance, good taste and excellence', and entered their 'English Sociable' to the Paris Exhibition where it was praised for 'uniting very easy action with lightness of construction and durability'.

Windscreen Wipers

One rainy night in 1917, J.R. Oishei was driving his car in Buffalo, USA. His vision was impaired and he knocked down a cyclist. The collision had a rather more lasting impact on Oishei than the cyclist who walked away, relatively unharmed. Oishei described the incident as 'a harrowing experience which imprinted on my mind the definite need for maintaining vision while driving in the rain'. He founded the Tri-Continental Corporation (Trico) and began making windscreen wipers. They opened in the UK in 1928 on the Great West Road, Brentford. The plant moved to Pontypool in the 1990s.

Royal Mint

The right of coinage has always been a royal prerogative. In 1275, William de Turnemire was appointed master-moneyer in all England and coining was then confined to the London Mint. Coins were minted at this time in the Treasury and Exchequer buildings at the Palace of Westminster, but then, in 1300, the Mint was transferred to the Tower of London. Between 1699 and 1727, the eminent scientist Sir Isaac Newton was Master of the Royal Mint, evidently with great success: 'The ability, the industry and the strict uprightness of the great philosopher speedily produced a complete revolution throughout the department which was under his direction.'

Advances in machinery and lack of space at the Tower prompted a move across the road to a building in East Smithfield designed by the surveyor to the Mint, James Johnson, and completed by his successor, Sir Robert Smirke. Opened in 1810, Johnson's Palladian building has a large central pediment with the royal coat of arms and six Doric columns beneath. East Smithfield closed in 1975, and coins are now minted in Llantrisant, South Wales.

Dog Food

James Spratt was born in Ohio and worked as a lightning rod salesman. He found himself in Liverpool in the mid-nineteenth century and at the dockside observed stray dogs eating ship hardtack. So inspired, Spratt patented a dog biscuit which he called the 'Patented Meat Fibrine Dog Cake'. He set up in Holborn and got the firm of Walker, Harrison & Garthwaite to make them. His first factory was in Poplar on the banks of the Limehouse Cut, near Morris Road. Ingredients were blended wheat meal, vegetables, beetroot and meat. He employed a young man called Charles Cruft as clerk, who went on to greater things when he founded Cruft's Dog Show. Spratt's ingredients were offloaded in Limehouse Basin and conveyed along Limehouse Cut to the factory. There they were blended into a dough, drawn through a machine and rolled to the required thickness. Then 50,000 biscuits per hour were cut out and placed in a vast oven, emerging 'row upon row of brown, healthy fellows, each and every one of them done to a turn'. Spratt's claimed to have supplied 1,256,976,708 biscuits to army dogs in the First World War.

The animal kingdom was also catered for by 'Sherley's Foods and Medicines for Dogs and Cats' with premises at 18 Marshalsea Road, Southwark. Their brochure

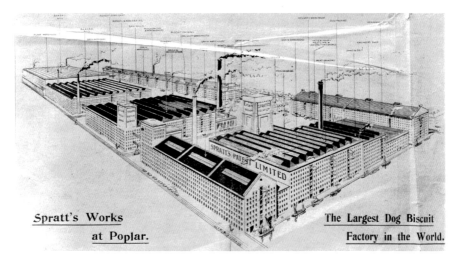

Spratt's. *(By permission of Tower Hamlets Local History Library)*

stated that some items contained a small percentage of an ingredient mentioned in the Poisons Act but was quite harmless if used in accordance with the instructions on the packet. One such was 'Sherley's Aphrodisties' – 'a cure for impotence in dogs and cats and invaluable for shy breeders': price 1s 6d, postage 1d.

False Teeth

The factory of Claudius Ash, Manufacturer of Mineral Teeth, was in Anglers Lane, Kentish Town. It was built in 1864, but the firm had been founded in 1825. At that time, Ash was trading as a silversmith in St James's Street and later in Broadwick Street, Soho. Over time he diversified into making false teeth of gold and silver (an alternative to taking teeth from corpses) and this soon became his main business. By the 1880s, Ash had an international reputation, with overseas branches in Berlin, Paris, Copenhagen, St Petersburg and Vienna.

Rubber Industry

Thomas Hancock, born in Marlborough in 1786, is known as the founder of the British rubber industry. In 1815, he was in London, working as a coach builder. His interest in rubber was aroused by his wish to protect the passengers riding on his coaches from the worst effects of the weather. Hancock invented a machine to shred waste rubber, calling it his 'pickling machine'. It was in fact a masticator but Hancock – to preserve its secrecy – preferred the anonymity of his own description. Hancock's first factory was in Goswell Road but some years later his brothers, Charles and Walter, set up the Gutta Percha Co. in High Street, Stratford. They were joined by Henry Bewley who relocated the firm to Wharf Road, by the Regent's Canal, and split from the Hancocks, who, at the same time, founded a competitor firm, the West Ham Gutta Percha Co. in Abbey Road.

Meanwhile, Stephen Winckworth Silver opened a factory in what became known as Silvertown. In 1864, the factory became the India Rubber, Gutta Percha & Telegraph Co. and held patents for making waterproof and insulating materials. It also took over Hancock's West Ham Gutta Percha Co. Throughout the nineteenth century the firm specialised in the manufacture of submarine cables. By the beginning of the twentieth century it was making bicycle and motor car tyres, and by 1923, on a 17-acre riverside site, employed 4,000 workers. The factory, now acquired by the British Tyre & Rubber Co., was badly bombed in the Second World War and in 1955 was renamed the Silvertown Rubber Co. Rubber production ceased in Silvertown in the 1960s.

Crocketts' Leather Cloth factory occupied the site where previously stood the West Ham Gutta Percha Co. in Abbey Road. In the 1930s, they employed 500 workers.

Tyre Manufacturing

Harvey S. Firestone was born in Chicago and in 1900, at the age of 31, left for Akron, Ohio, the centre of the US rubber industry. He began making solid rubber tyres for bicycles and carriages, and opened a factory in England in 1928 at Brentford on the Great West Road, a location he chose because of its good railway links and access to the Docks via the Regent's Canal. Firestone's certainly impressed the Home Secretary, Sir William Joynson-Hicks, when he came to inspect it. There was a spacious grass lawn in front and Joynson-Hicks commented that 'working conditions were excellent and he wished British manufacturers had a similar regard for the health and comfort of labour'. The building, in the Art Deco style, was designed by Wallis, Gilbert & Partners. Much to the distress of those appreciative of fine buildings, it was demolished in 1980, one day before a preservation order was due to be placed on it.

Hat Making

In the sixteenth century there was a change in the fashion of men's headgear. They began wearing felt hats instead of woollen caps. At the same time, French and Dutch felt makers settled in Bermondsey and began making felt hats. So worried were cap makers, that it became an offence for certain groups of men *not* to wear a cap! Despite all regulations, more and more began wearing the newly fashionable headgear. The hair necessary for the manufacture of felt was a by-product of Bermondsey's leather industry. By the end of the century, the industry was well established, as evidenced by the 'great disturbance' which occurred amongst hatters when one of them was unjustly locked up in the Marshalsea

Prison. In the nineteenth century, over 3,000 were working in Bermondsey, gaining for the parish the title 'Hatters' Paradise'.

Messrs Christy's & Co. were a prominent firm in Bermondsey occupying two extensive ranges of buildings in Bermondsey Street. They began in Gracechurch Street in 1773, and also had a factory in Stockport. The Bermondsey branch specialised in silk top hats but also made policemen's helmets, undergraduates' mortar boards, hats for Beefeaters at the Tower of London, Chelsea pensioners' hats and gold-embroidered cocked hats for doorkeepers at the Bank of England. It was their boast that they employed 'more men by half' than any other hat manufacturer in the country.

George Carter & Sons were founded in the Old Kent Road in 1851 and made silk top hats. Their trademark was an upturned hat floating in water and surrounded by cherubs, with the motto 'Carter's Noted Hats Light and Waterproof'. Another top hat was outside the shop, perched on a pole. Just the same as the timepiece at the Royal Observatory at Greenwich, it fell at exactly 1 p.m. every day. Later, the firm had a clock with a hat that was raised on the hour.

Lock & Co. of St James's were (and are) London's most prestigious hat makers. It was their custom to name any particular hat after the first customer who bought one. A hard fur felt hat with a semi-rough finish was sold to William Coke,

Lock & Co., St James's.

a Norfolk farmer, but in fact the hat had been manufactured by Thomas Bowler who had premises in Southwark Bridge Road. Bowler was born in Stockport into a family of hat makers and moved to London in 1837. He went into partnership with a Frenchman, Victor Jay, and in 1850 made the first ever Bowler hat (or Coke hat). Some called it the 'iron hat' because of its strength; it was so hard that Mr Coke successfully tested it by jumping on it. In America it is called the 'Derby', a reference to its first appearance at the Epsom Derby. So popular was the bowler that 60,000 were made every year and it became the trademark of the City gent. King Amanullah of Afghanistan visited London in 1936 and was so taken by his bowler-hatted security guards that he ordered a gross of hats to take back with him in anticipation of substituting them in place of the traditional fez. But he forgot that Muslims at prayer are required to touch their heads on the ground – an impossible task with a bowler. The fez remained supreme! The bowler hat celebrated its centenary in 1950 and to mark the occasion a grand party was held at London's Dorchester Hotel.

Mathematical & Scientific Instrument Makers

Cooke, Troughton & Simms started in 1688 when John Worgan set up shop under the dial of St Dunstan's Church, Fleet Street. By 1716, the shop was run by John Rowley whose other business was at Johnson's Court, at the sign of 'The Globe'. The firm had a prolific output, making all sorts of mathematical instruments – pocket dials, compasses, protractors, scales, orreries and sextants, including astronomical instruments for Trinity College, Cambridge, which included telescopes and globes for the great Sir Isaac Newton. Later owners of the firm at the 'Orrery and Globe', Fleet Street, were Thomas Wright and Benjamin Cole who became mathematical instrument makers to George II. In 1782, Edward and John Troughton took over from Benjamin Cole and later joined forces with William Simms. It was at this time, soon after the Imperial Standards of Length were destroyed when the Palace of Westminster burned down in 1834, that the firm made the standard yard which is situated in Trafalgar Square. In 1864, a move was made to Woolwich Road, Charlton. The firm merged with T. Cooke of York in 1922.

The Italian Henry Negretti was born near Lake Como and travelled to London via the St Gotthard Pass to begin his business in 1843 in Old Leather Lane. In 1850, he joined with Joseph Warren Zambra to found a scientific instrument company, Negretti & Zambra, which was soon to have a worldwide reputation. The firm made meteorological instruments, including a maximum thermometer and later a double-bulb deep-sea thermometer for taking the temperature of the ocean at great depths. A mercurial barometer was developed for use aboard fighting vessels. It was able to withstand the vibrations caused by the discharge of guns. Aneroid barometers, telescopes and rain gauges were also made. The firm's premises were in Hatton Garden and later at Holborn Viaduct. After the First

World War, Negretti & Zambra concentrated on making industrial and aeronautical instruments. They moved from Holborn Viaduct in 1964 to open a factory in Aylesbury.

As well as being a skilled instrument manufacturer, Negretti was something of a detective. On Boxing Day, 1865, tempers became frayed at the 'Golden Anchor' public house and a brawl started between groups of Italians and local men. An Englishman was fatally wounded and blame was levelled at an Italian, Serafino Pelizzioni. The offending weapon was found some distance from the pub and Pelizzioni was duly charged with murder, convicted and sentenced to be hanged. Negretti became interested in the case. He was well known and respected in the Italian community and was puzzled by the fact that the assailant had been apprehended in the pub and so did not have the opportunity to dispose of the weapon outside. Furthermore, it appeared that Pelizzioni had been drinking in another pub at the time of the murder and that the dirty deed had in fact been perpetrated by his cousin, Gregorio Mogni. Negretti was successful in bringing a private prosecution, with the result that Mogni was convicted of manslaughter and Queen Victoria granted Pelizzioni a full pardon.

Part 3

TRANSPORT

21 Canals

Until the coming of the railway era, goods were more easily transported by water; roads were little more than rutted tracks and merchants from very early on used either rivers or the sea. The Industrial Revolution gave canal building an enormous boost, and the first canal dug specifically for industrial purposes was the Sankey Canal in 1757, built to carry coal from the Lancashire coalfields to Liverpool. It was followed by the Bridgewater Canal in 1761, also for transporting coal.

Birmingham and the West Midlands area saw a vast increase in its manufacturing industry at this time. It was hampered, however, by difficulties in transporting goods to markets in London and abroad. The Oxford Canal went some way to alleviating the problem. It was promoted by Sir Roger Newdigate – landowner, collector of antiques, philanthropist and MP for Oxford University – and was authorised by Act of Parliament in 1769. The work of James Brindley and, after his death in 1772, Samuel Simcock, the canal was completed in 1790 and ran from Coventry (with its links to Birmingham) to Oxford and the Thames. Goods – typically coal and stone – were then transported to London along the River Thames. But navigation by river was far from ideal, particularly in the upper reaches of the Thames which were too shallow. Proposals were therefore made for a direct canal link from the Midlands to London and the south. Two schemes were put forward. The first was to take a branch from the Oxford Canal at Thrupp and route it to Marylebone via Aylesbury, Wendover and Uxbridge. But it was the second plan – the Grand Junction scheme – that won the day.

The Grand Junction Canal

The Grand Junction was an extension of the Oxford Canal at Braunston, Northamptonshire, to link with Brentford and the Thames via Weedon, Wolverton, Leighton Buzzard, Berkhamsted, Watford, Harrow and Greenford. The Act came onto the statute book on 30 April 1793, and the company lost no time in holding the first meeting of the board on 1 June 1793 at the 'Crown and Anchor Tavern' in the Strand. William Praed was appointed chairman and James Barnes chief engineer. Barnes was paid 2 guineas a day and was helped by

Junction of the Paddington Branch of the Grand Junction Canal with the Grand Union Canal at Bulls Bridge, Southall.

William Jessop, who acted as consultant. The Grand Junction Canal was dug with great haste – 3,000 men were employed every day – and within a year it was decided that as well as linking with the Thames at Brentford, an extension should be dug to a place nearer the centre of London. Accordingly, an arm was dug to Paddington in 1797 from Bulls Bridge in Southall. (There were even plans to extend further towards the City, to Tottenham Court Road, but they did not come to fruition.) On 10 July 1801, the 60ft-wide extension ending at Paddington Basin was opened, thus reducing the distance from London to the Midlands by 60 miles. The canal was a great success, carrying over 340,000 tons a year by 1810.

The desirability of linking the River Lea with central London was recognised from early days: in 1571, an Act proposed 'bryngyng of ye Ryver of Lea to ye Northside of ye Citie of London'. Its route would have been from Temple Mills on the Lea to Moorgate via Dalston and Haggerston. There were further schemes in 1770, one of which was to join the Lea with the ill-fated London to Birmingham scheme at Marylebone. Neither got off the ground.

The Regent's Canal

A canal to link the Thames at Limehouse with the Grand Junction at Paddington was the brainchild of Thomas Homer in the early nineteenth century. He asked John Rennie to draw up plans, but these took a route through central London and proved to be too costly. Homer was not discouraged by earlier failures and resurrected his idea in 1810, this time for a route around the northern limits of London. It met with a measure of approval, and the first meeting of the canal's proprietors was held at the 'Freemasons' Tavern', St Giles in the Fields, with Homer appointed as superintendent at a salary of £400 per year.

As expected, there were the usual objections. An 'Observer' wrote a pamphlet complaining of 'the permanent interruption to be occasioned by 29 public bridges, rendering useless 72 acres of land, introducing bargemen and others into land otherwise private and the insecurity to the public from persons passing through a line of country for 9 miles at all times of night'. Objections or not, the scheme went ahead.

Built in collaboration with James Morgan and John Nash (who got the Prince Regent to give his name to the canal), the scheme was the work of the Regent's Canal & Dock Co. It soon hit problems: there was a tragic accident near Chalk Farm where a bank collapsed and buried twelve men. Eight were rescued alive and severely injured but four were dead when dug out. To add to the proprietor's troubles, Thomas Homer was found to have embezzled large amounts of the company's money. He fled to Ostend and then to Scotland but was finally caught up with and deported. Nevertheless, on the Prince Regent's 54th birthday, 12 August 1816, a 10-mile stretch was opened from the Grand Junction Canal at Paddington to Hampstead Road. The canal was finally completed from Paddington to Limehouse in 1820, and on 1 August of that year, to celebrate, the proprietors and their guests were conveyed with much splendour on state barges to Limehouse, passing en route beneath Islington through the tunnel before retiring to the 'City of London Tavern' for a fine dinner. Trade was brisk; in 1821, there were 195,000 tons carried, increasing to 470,000 tons in 1830.

The Limehouse Cut had been completed some years before, by Thomas Yeoman in 1770. It now branches off from Limehouse Basin but originally left the Thames a few yards to the east to connect with the Lea Navigation at Bromley-by-Bow, thus allowing barges to avoid the long journey around the Isle of Dogs.

In 1829, the Regent's Canal Co. combined with the Grand Junction Canal Co. to form the Grand Union Canal Co. Trade increased and included the import of the raw materials used to make HP Sauce in Birmingham.

Limehouse Basin was originally intended for barges but was later increased in size to take ships, including the new iron colliers. By the second half of the twentieth century, commercial traffic was in terminal decline. The canal is now used for leisure purposes, including the pleasure boat service 'Jason's Trip', which runs from Regent's Park.

The Grand Surrey Canal

The first canal south of the river was the Grand Surrey Canal – that is, if we disregard Cnut's canal. The romantic story of Cnut's canal is worth recalling. In 1016, the Saxon Edmund Ironside held London north of the river. His rival, the Viking Cnut, who was camped at Greenwich, allegedly built a canal from there to Battersea to outmanoeuvre him. A fantastic story! Much later in 1790, a Mr Cracklow suggested another scheme for a canal to link Deptford with Bankside. It was not built.

The Grand Surrey Canal was the idea of Ralph Dodd, an engineer, who in 1801 obtained powers to construct a canal from Wilkinson's Gun Wharf on the riverside at Rotherhithe to Mitcham in Surrey. Dodd, author of *An Account of the Principal Canals in the Known World*, was a man with grand ideas. He once had an ill-fated plan to build a tunnel under the Thames between Gravesend and Tilbury. And his plan for the Grand Surrey Canal was similarly ambitious. He proposed that it should extend to Epsom in Surrey, and even to Portsmouth and the coast. In the event, the proprietors lost interest in Dodd because of his high costs and the project was put out to tender. John Dyson came in with the lowest bid, which was accepted, and Ralph Walker was appointed as engineer. The canal ran for 3½ miles from the Thames at Rotherhithe, across the Greenland Dock and through the market gardens of Deptford and Peckham to Camberwell. It was the first canal to have a police force: Bank Rangers, as they were called, were appointed to keep order. It was the proprietors' original intention that the canal would serve to transport market gardeners' produce to the people of London, but the gardens were soon to be swallowed up by the ever-expanding thrust of London itself.

The canal opened, with the usual ceremony, on 13 March 1807:

> … in the presence of a numerous assemblage of spectators, composed principally of the proprietors and their friends, together with a large company of ladies, who all appeared much gratified on this interesting occasion … guns were fired as a signal for the first vessel to enter and at 3 o'clock the Argo, dressed in the colours of various nations, entered amidst the acclamations of the spectators. She was saluted by a discharge of cannon from the shore which was returned by the vessel; while the band of martial music on the deck played 'God Save the King' and 'Rule Britannia'. The whole made a very interesting appearance.

The South Metropolitan Gas Co. made good use of the canal, moving their coal along it to their Old Kent Road works.

The Grand Surrey Docks & Canal Co. built many of the timber storage ponds which lay between the Greenland Dock and the river to the north, and these proved to be more profitable than the canal itself. The ponds were eventually filled in when the Surrey Docks closed. The canal itself closed in 1971.

The River Lea Navigation

The River Lea rises near Luton, and it is said that the Romans used it to enable them to reach their important city of Verulamium (St Albans) from the Thames.

In AD 896, Alfred the Great, together with a fleet of his ships, is said to have pursued the Danes along it. The area has been heavily industrialised since the mid-nineteenth century and its pollution was summed up in *Punch* in 1885 with a poem called 'The Rowers on the Lea':

> Within a brief half hour,
> They sang but not in glee,
> We envy folk upon the bank,
> But they don't envy we!
> For why? We feel inclined to faint,
> We're sick as sick can be;
> We've all got germs of typhoid from
> This rowing on the Lea.

As well as this there was:

> There are ten bad smells of Lea,
> They are vile as vile can be
> And there they are, and there they'll be,
> Unless to the matter the public see,
> Those ten bad smells of Lea.

The plight of the River Lea was a direct consequence of the Metropolitan Buildings Act of 1844 – which sought to protect populated areas from the new noxious-smelling industries of the nineteenth century. The area of Plaistow on the banks of the Lea fitted the bill ideally!

The first record of any commercial traffic on the River Lea was in the thirteenth century when Margaret, Countess of Winchester gave the canons of Holy Trinity right of passage for their corn from Ware to London for 1*d* per quarter. There were many water mills along its length, including Three Mills at Bow, which still survives today, although not in use.

In 1766, the River Lea Act was passed and new locks and cuts were opened; others were cut after a further Act of 1850. By this time, and certainly within the Lower Lea Valley, the river was effectively a canal and heavily polluted. The Lea and its associated waterways tended to flood, particularly in the Stratford area, and as a result, the River Lea (Flood Relief) Act of 1930 came into force, enabling extra channels and locks to be constructed. The whole complex of waterways became known as the Bow Back Rivers.

There is now a bright future for the Lea. The London 2012 Olympic Games has given an enormous boost to the river and canal by measures taken to prevent flooding and by the installation of new locks to render the river non-tidal. The river runs through the site of the Olympics, which in 2013 is to be transformed into the Queen Elizabeth Olympic Park for recreational use.

Other Canals

In 1799, the ubiquitous Ralph Dodd proposed a canal – the Croydon Canal – to link Croydon with the Thames at Deptford. John Rennie suggested another scheme to end at Rotherhithe. Both were overtaken by events with the passing of the Grand Surrey Canal Act and accordingly, plans were made for the Croydon Canal to join with the Grand Surrey and thence to the Thames. The canal had a chequered history: canal mania was overtaken by railway mania and, in 1836, the London & Croydon Railway Co. purchased the canal. They laid their railway line along most of its length. West Croydon Station now stands on the site of the canal basin.

In another part of London, proposals were put to William Jessop by a group of businessmen for a canal to link Wandsworth and the Thames with Croydon. Jessop was not impressed, concluded it was impractical and suggested a railway instead – the Surrey Iron Railway. He did propose, however, a canal basin at Wandsworth to aid the transfer of goods between river and railway. The basin was known as MacMurray's Canal and, in 1802, 'the first barge entered the lock amidst a vast number of spectators who rejoiced at the completion of the important and useful work'. It lasted for over a century but after the First World War the canal basin fell into dereliction.

The Kensington Canal was conceived by Act of Parliament in 1824. Construction was put out to tender and a Mr Hoof got the job. It ran from just west of Battersea Bridge, through open countryside, and terminated at the Great West Road, about ½ mile west of Kensington Palace. The canal opened in 1838 and the proprietors charged 3½d per ton to carry manure and 7d for other goods. Plans were then laid to extend the canal to the Grand Junction at Paddington, but were

MacMurray's Canal, Wandsworth.

not realised. However, the Birmingham, Bristol & Thames Junction Railway Co. came on the scene with proposals for a railway line from the canal basin (at Warwick Road, just north of present-day Cromwell Road) to link with the London & Birmingham Railway. The canal was finally filled in, apart from a short stretch from the Thames to King's Road, Chelsea. But it did not die and after the First World War it positively thrived, carrying imports of coal to the Imperial Gas Works and removing waste from the works. Commercial traffic finally ceased in 1967 when the barge *Rabbie* delivered the last shipment of 23,000 gallons of oil to the gasworks.

The Grosvenor Canal began as a small cut into the Thames dug by the Chelsea Water Works in 1727 for the purpose of extracting and supplying drinking water. The land belonged to Lord Grosvenor, who had ambitions for a proper canal carrying commercial traffic. Accordingly, a contract was signed with John and William Johnson and Alexander Brice in 1823, and work began to extend the cut inland. The canal opened in 1825, carrying on uneventfully until the railway intervened. In 1858, an Act of Parliament authorised 'stopping up or appropriating the Grosvenor Basin and using some of it for the purposes of the railway station and works' – today's Victoria Station. A small section of canal continued and after the Second World War was used for refuse removal.

22 The Docks

It all started with the Romans. Remains of a Roman harbour were found in the 1820s when London's medieval bridge was demolished. Later excavations around London Bridge revealed a wooden quay west of the bridge, dating from around AD 70, and one east of the bridge built twenty years later. By AD 410 the Romans had gone – now came the Saxons.

The early Saxons were not a sophisticated people. Their needs were food, shelter and warmth. Furthermore, they were not interested in living in towns, preferring instead to live in small farming communities. So Roman London, including its port, fell into ruin and probably stayed that way for 200 years. The Saxons settled just to the west of the City walls and established a community known as Lundenwic. We are informed by the Venerable Bede, writing in AD 731, that Lundenwic was 'a mart of many nations resorting to it by sea and land'. But there was no port in the sense we understand today. Gustav Milne in *The Port of Roman London* tells us the mart would have been a very small-scale affair, with trade carried out at beach markets at the riverside.

After the Norman Conquest, the waterfront began to be gradually improved by wooden embankments at Billingsgate, Dowgate and Vintry. There were two important hithes (artificial inlets) at Queenhithe and Billingsgate. Queenhithe was originally known as Ethelredshythe, named after King Alfred's son-in-law, and is referred to in a charter of AD 899. It later took the name Queenhithe when given to Queen Matilda by her husband, Henry I.

Despite the impediment of having to navigate London Bridge, goods were landed preferentially at Queenhithe. In 1226, Henry III commanded the Constable of the Tower of London to compel ships of the Cinque Ports to land their corn at Queenhithe and that all fish was to be confiscated if not sold there. Furthermore, if any foreign vessel unloaded fish at Billingsgate it should be fined 40s.

The reign of Elizabeth I saw the port and trade flourish. London's late medieval harbour stretched along the entire length of the City's waterfront. It was lined with wharves which often took their owner's name or that of the type of cargo handled there. Most goods were subject to Customs duty. Corruption was rife and early in the reign of Elizabeth I it became apparent that revenues were falling seriously short because of 'greedy persons', smuggling and corrupt Customs officials. Therefore, in 1558, an Act of Parliament was passed which decreed that all dutiable goods should henceforth be handled only during the hours of daylight and at specially designated quays, which came to be known as the Legal Quays. They were situated along the north bank of the river between London Bridge and the Tower.

Wool Quay was long the home of the Customs House. It was the responsibility of the Customs House to collect both import and export duty. A Customs official, known as a 'Tide Waiter', would board an incoming ship at Gravesend and sail with her to her mooring place at the Legal Quays. The vessel's captain registered his cargo in the Long Room at the Customs House and then a 'Landing Waiter' would assess the ship's cargo. Not surprisingly, it was a complicated business and open to all kinds of abuse.

In 1661, a large wet dock of 1½ acres was built at Blackwall, the first wet dock on the Thames, and at the end of the seventeenth century the Howland Dock was constructed at Rotherhithe. It must be emphasised, however, that both were restricted to the repair and fitting-out of ships and not the handling of cargo. As the eighteenth century progressed, trade flourished at an ever-increasing rate. But as trade increased so did the theft of cargo, and it was against this background that people began to argue for a system of enclosed docks. Chief amongst these were the West India merchants who were losing as much as £500,000 per year from theft. The government also lost money because of unpaid Customs duty. Two men emerged to give voice to the general discontent. They were the Dumbarton-born and former Lord Provost of Glasgow, Patrick Colquhoun, and the London merchant William Vaughan, both of whom were foremost in urging changes in the way the port operated.

To combat theft, Colquhoun campaigned vigorously for a river police force and wrote on the subject in his book, *A Treatise on the Commerce and Police of the River Thames*. All dutiable goods, of course, had to be offloaded at the Legal Quays, still situated between London Bridge and the Tower, and offering no more landing room than in Elizabethan days. In an attempt to relieve congestion, a series of Sufferance wharves had been established for goods of lower value by an Act of

1663. Five were on the north bank between the Tower and Hermitage in Wapping, and eighteen on the Southwark side, giving a total length of 3,676ft. Taken together, the total quay space was only a little over 5,000ft, totally inadequate for the demands put upon it. The most significant import from the West Indies was sugar. Ocean-going vessels would anchor downstream and transfer their cargo to lighters. At any one time, as many as 200 lighters would remain laden with sugar, waiting to gain access to the Legal Quays. Once there, their problems were far from over, for warehousing space was inadequate: only 32,000 hogsheads of sugar could be accommodated at any one time. So, in effect, the lighters themselves often acted as temporary warehouses and were consequently open to theft. Colquhoun's campaign was successful. A river police force was established and fortress-like high walls surrounded the soon-to-be-built new docks.

Real change in the way the port operated came about because of the endeavours of William Vaughan and his pamphlet *On Wet Docks, Quays and Warehouses, for the Port of London; with Hints Respecting Trade*. He told how:

> … the West India trade has been for years labouring under the severest burthens from delays, charges, losses and plunderage … it is therefore necessary from increased imports and growing impediments to commerce in all its branches, to apply some remedy, and none can be more effectual than the creation of docks and quays, with an extension of warehouses.

It was obvious that something had to be done, so to address the problem a committee of the House of Commons was set up with the brief of inquiring into the best mode of providing accommodation for the increased trade and shipping of the Port of London. Two separate schemes were therefore put before Parliament in the 1799 session. The Isle of Dogs scheme came to the statute book first, on 12 July 1799, followed by the Wapping scheme, on 23 May 1800. So were born the West India Dock at the Isle of Dogs and the London Dock at Wapping.

West India Dock

The West India Dock Act of 1799 provided for two parallel docks across the Isle of Dogs and a ship canal, the City Canal, to enable a shortcut between Blackwall and Limehouse to avoid the long journey around the Isle of Dogs.

An inscribed plaque, still there today, describes the laying of the foundation stone on 12 July 1800. It was a grand occasion and the excuse for a public holiday. The stone was laid by the Lord Chancellor, Lord Loughborough, in the presence of many dignitaries, including William Pitt the Younger. The plaque reports that it was all 'at the expense of public spirited Individuals' and 'under sanction of a provident legislature'. Present also were the chairman and vice-chairman of the West India Dock Co., George Hibbert and Robert Milligan. The chief engineer was William Jessop who had previously built the Grand Union Canal, the vital

West India Import Dock, 1810. *(By permission of Tower Hamlets Local History Library)*

trade link from London to the Midlands. Ralph Walker was the resident engineer and John Rennie acted as consultant. Later, in 1802, in the presence of Prime Minister Henry Addington, the dock was officially opened when a ship taking his name, and bedecked with the flags of all nations, sailed into the dock in some style. But it was soon down to business and immediately afterwards the *Echo* arrived laden with the first consignment of sugar.

West India Dock was London's first enclosed dock constructed specifically for handling cargo. The dock was something of a fortress, surrounded by a 30ft-high brick wall, on the outside of which was a ditch 12ft wide and 6ft deep. Inside there were sentry boxes and, in the early days, troops. All this was to stop the rampant pilfering that Colquhoun had written about so vividly and to ensure that Customs revenues were not compromised. In place of the Legal Quays, a series of bonded warehouses were constructed by the father and son team of architects, the Georges Gwilt. Two of these fine warehouses remain to this day, one housing the Museum of Docklands. They were secured by locks kept jointly by the Inland Revenue and the dock company. Under this new arrangement the Revenue allowed payment of dues when the cargo – primarily sugar – left the warehouse for the outside world. As other docks were built, similar arrangements were put in place. The West India Dock had a twenty-one-year monopoly on the handling of goods in all ships sailing to and from the West Indies, the only exception being tobacco. When the West India Dock opened, vessels had to pay 6s 8d per ton and there were also charges for cargo. Sugar, for instance, was charged at 8d per hundredweight. The West India Dock Co. soon took over the City Canal, which ran between Blackwall Reach and Limehouse Reach, south of the West India Export Dock. The canal had turned out to be a white elephant – the inconvenience of vessels having to pass through two locks resulted in little, if any, time saving. It was the dock company's intention to convert the canal to a dock, the South West India Dock, a project they eventually completed in the 1870s.

London Dock

The London Dock Co. received royal assent in 1800 and, despite objections from the owners of the Legal Quays, watermen, lightermen and, of course, the West India Dock Co., work soon began. The foundation stone was laid on 26 June 1802, when the 'docks were crowded with genteel persons of both sexes'. Daniel Asher Alexander was chief surveyor and architect. Born in London and a student at the Royal Academy, Alexander was a fine architect, much praised by Sir John Soane for his work at Inigo Jones' Queen's House at Greenwich. Also surveyor for Trinity House and responsible for many well-known lighthouses, including on Lundy Island, he put his experience of building secure docks with high walls to good effect as architect of Dartmoor and Maidstone prisons. He was much praised by *The Gentleman's Magazine*, which wrote of his work: 'A characteristic fitness of purpose was prominent in every building.' It was, however, John Rennie who did much of the work at London Dock.

The first ship sailed into London Dock on 31 January 1805 at Wapping Pier Head, and the company enjoyed a monopoly lasting for twenty-one years on all trade in wine, rice, tobacco and brandy, except that to and from the West and East Indies. London Dock closed in 1969 and has now largely been filled to make way for the giant News International premises, otherwise known as 'Fortress Wapping'.

Opened as the New Tobacco Warehouse in 1814, and a Grade I listed building, Daniel Asher Alexander's marvellous building, with its roof of timber and cast iron, was unique at the time and still survives today. The building covered a vast area of 80,000 square feet, with brick vaults beneath. It later became known as the Skin Floor and from 1860 concentrated on the import of wool, tobacco moving

London Dock, by William Daniel, 1808. *(By permission of Tower Hamlets Local History Library)*

to the giant new Victoria Dock at Plaistow. Sheepskins, with the wool still intact, were imported from Australia, New Zealand and the Falkland Islands, and were stacked in huge bales for inspection and later auction at the Wool Exchange in Coleman Street in the City. Wines and spirits were stored in the vaults below.

Shadwell Basin was the main entrance to London Dock. The basin has survived and is now surrounded by pleasant and original new houses. They have galleries, porthole windows, arcades and bright colours to reflect the area's Docklands past.

East India Dock

The East India Dock was originally associated with the East India Company. As mentioned above, a large wet dock of 1½ acres was built at Blackwall Yard in 1661, the first wet dock on the Thames, for the fitting-out of vessels. Towards the end of the eighteenth century, the then owner, John Perry II, enlarged the dock to 8 acres and it became known as the Brunswick Dock. Then, between 1803 and 1806, John Rennie and Ralph Walker constructed the East India Dock; this included two parallel docks for both export and import trade, and the export dock was built on the former Brunswick Dock. It was London's third set of wet docks. The docks had a twenty-one-year monopoly on trade from China and the East Indies. In the early days, and in contrast to the West India Dock, the East India Dock was not lined with warehouses; instead, the East India Company unloaded their cargoes into covered wagons to be taken via the newly opened Commercial Road to their splendid warehouses in Cutler Street in the City.

The East India Dock opened in August 1806 with a splendid ceremony similar to that seen a few years before at the West India Dock. The *Admiral Gardner* was

Brunswick Dock at Blackwall, 1803, by William Daniel. *(By permission of Tower Hamlets Local History Library)*

East India Dock, by William Daniel, 1808. *(By permission of Tower Hamlets Local History Library)*

first to enter the dock, flying the flags of all nations, with, pointedly, the French flag beneath all others (the Battle of Trafalgar had only recently been fought and won). On the ship's quarterdeck was the band of the East India Company, playing 'Rule Britannia' with a full chorus sung from all decks. Afterwards, all dignitaries retired for 'an elegant dinner' at the 'London Tavern' – an opulent eating place in Bishopsgate famed for its turtle soup. During the Second World War, important work was carried out in the import dock. It was emptied and used for the construction of concrete caissons, the so-called Mulberry Harbours, to be used as prefabricated ports for the D-Day landings in Normandy.

Greenland Dock

Greenland Dock, situated between Rotherhithe and Deptford, was opened in 1699 as the Howland Great Wet Dock. It is the oldest surviving wet dock in London's dock system and was much smaller when first built than it is now. It was extended some 1,070ft from the river bank and was capable of holding 120 of the largest merchant vessels. In its early days the dock was used solely for the fitting-out of vessels and as a safe haven. It is highly probable that it was built by John Wells, a member of a famous shipbuilding family. In fact, in 1763, the dock, which had by then become known as the Greenland Dock (so named because of the whaling trade which was centred on the quayside), passed to the Wells concern. Then, in 1806, it was purchased by the Commercial Dock Co. In 1808, they appointed James Walker as their chief engineer and he soon set about rebuilding.

The Commercial Dock Co. amalgamated with the Grand Surrey Docks & Canal Co. in 1865 to form the Surrey Commercial Docks Co. Timber was their main trade and the company also dealt in grain. But it was soon to face competition from the newly opened Millwall Dock and the Royal Docks on the

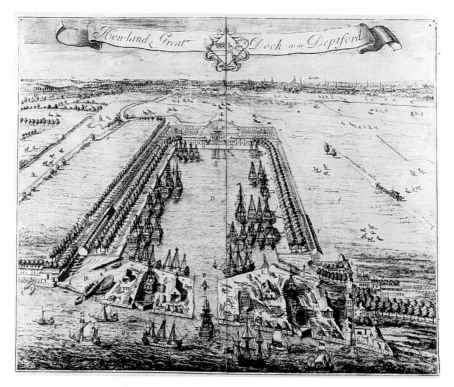

The Howland Great Wet Dock, Rotherhithe, *c.* 1717. *(By permission of Southwark Local History Library)*

other side of the river, both of which could hold larger vessels. To counteract the threat, the company determined to extend and enlarge the Greenland Dock. Accordingly, in 1894, work started under James MacConnachie and was later continued by Sir John Wolfe Barry. The dock was extended to its present length at a cost of £940,000. Other docks were built by the company to the north and by 1921, there were nine docks and six timber ponds in the whole complex. The Surrey Docks suffered greatly during the Second World War and, in common with the docks in general, were a prime target for German bombers. One dreadful night in September 1940 saw 350,000 tons of timber destroyed in what was the largest fire ever in Britain.

St Katharine Dock

Now a marina, offices, shops and flats, St Katharine Dock was built in 1828 on the 23-acre site where once stood the Hospital of St Katharine by the Tower, a brewery and 1,100 houses. In the mid-1820s, the St Katharine Dock Co. was formed with the intention of building a dock close to the City of London. The directors were spurred on by the imminent expiry of the monopolies held by the West India, London, and East India Dock companies, and based their case on the

The opening of St Katharine Dock. *(By permission of Tower Hamlets Local History Library)*

ever-increasing shipping in the Port of London. And on the face of it, the company had a good case – in 1794, a total of 13,949 ships entered the port, but by
1824 this had increased to 23,618. Opposition to the plans was fierce, not so much
because people would be thrown out of their homes, but because St Katharine's
would be demolished together with its church, one of the few medieval churches
to survive the Great Fire. The docks took two years to build, employed 2,500
in the process and opened in 1828. They were the only example of the work of
Thomas Telford in London.

The warehouses were by Philip Hardwick and stored many commodities,
including tea, matches, marble, ivory and even live turtles. The Ivory House, built
in 1858 as the centre of London's ivory trade, has a distinctive Italianate tower by
George Aitchison, Hardwick's successor. Resistant to fire because of its wrought-
iron and brick construction, it is the only original building in the dock and was
unique at the time, with its walls built directly to the water. Telford's dock design
consisted of an entrance basin of 1½ acres leading to two docks, east and west.

Millwall Dock

Named from the windmills that used to line the wall on the west side of the
Isle of Dogs, Millwall Dock was opened by the Millwall Freehold Land & Dock
Co. on 14 March 1868. It was originally intended to provide water frontage for
manufacturers and shipbuilders, rather than engage in trade. It is L-shaped, and
covers some 36 acres, with its original entrance on the west side of the island.
It was designed by John Fowler and William Wilson. The company intended to
exploit the recent repeal of the Corn Laws, bringing with it the possibility of

cheap imports of foreign grain. Grain import, together with timber, thus became the dock's main business. The dock was lined with transit sheds (rather than warehouses for more lengthy storage) and was served by an extension of the Blackwall Railway – the Millwall Extension Railway – that ran through to the southern tip of the island to what was then known as North Greenwich. Millwall Dock was joined to the West India Dock by the Port of London Authority in the late 1920s.

The Royal Docks

In the mid-nineteenth century, the land which the Victory Dock was to stand on was acquired by three railway contractors – Samuel Morton Peto, Edward Ladd Betts and Thomas Brassey – who formed the North Woolwich Land Co. They were joined by the mathematical genius George Parker Bidder, also known as the 'calculating boy'. They got the land at what was termed at the time as an 'agricultural price' and it was an astute move because in 1844, the government passed the Metropolitan Buildings Act, which was to act as a signal for industrial development in the East End of London. Opened by Prince Albert, the Victoria Dock was intended from the beginning to link with the national railway network. As well as being served by the railway, it was the first dock in London to be constructed specifically for the new and larger iron steamships that were rapidly replacing the older wooden sailing vessels. It was equipped with hydraulic lifting machinery, newly introduced by Sir William Armstrong. All this gave the Victoria Dock a significant advantage over the older docks upriver. The entrance to the dock was from the west, but Peto, Betts and Brassey were astute businessmen, and while the Victoria Dock was being constructed, they were buying land to the east. And there to the east the Royal Albert Dock was built for the London & St Katharine Dock Co. (which had also taken over the Victoria Dock) by Sir Alexander Rendel. The dock is some 85 acres in area and was opened in 1880 by the Duke of Connaught on behalf of his mother, Queen Victoria. It was originally intended as a ship canal to give large ships access to the Victoria Dock, but then plans were changed and a complete dock was constructed. It was given royal designation from the beginning, and at the same time its near neighbour was promoted to the Royal Victoria Dock.

The two docks are vast in extent, and in the late nineteenth century were justifiably thought of as being at the 'hub of the Empire's trade importing every product known to man'. They specialised in chilled meat, tobacco and grain. There were electric lights from the start, and railway sidings on each side to enable the dock to be used for the transit of goods rather than long-term storage in warehouses. Plans to build another huge dock to the north were never realised, but in 1921, the Port of London Authority opened the 64-acre King George V Dock to the south, linked to the Royal Albert and with its own entrance lock to the Thames. On 8 July 1921, George V and Queen Mary arrived in great splendour by boat from central London, accompanied by the Duke and Duchess of York (the late Queen Mother). As the royal party arrived, children sang a selection of patriotic songs and there were gun salutes at the Tower of London and Woolwich Arsenal. The King opened the dock to a fanfare of trumpets in the presence of the chairman of the Port of London Authority and the Archbishop of Canterbury. Tea was served in one of the transit sheds and the party finally made its way back to London via the East End. So the Royal Docks were completed. They have a total area of 245 acres and 11 miles of quays, and at the time formed the largest area of wet docks in the world.

Postscript

Serious problems had become evident in the operation of the Port towards the end of the nineteenth century and in response the government set up a Royal Commission to investigate. It took evidence from interested parties, including the dock companies, the City Corporation and the London County Council, and concluded that a single authority was needed to run the entire Port. Royal assent was granted to the Port of London Act and the Port of London Authority (PLA) was duly founded in 1908. The PLA's powers were considerable. They took over the responsibilities of the Thames Conservancy as well as all the dock companies, but left the riverside wharves in private hands.

The docks were devastated by wartime bombing as they were obviously a prime target. Scores of warehouses were destroyed as well as other port property. As soon as hostilities ended, rebuilding began and a general mood of optimism prevailed. But old problems remained, in particular those concerned with industrial relations. What was to eventually sound the death knell for London's docks was the container vessel. Now cargo could be loaded into large metal containers measuring 8ft x 8ft x 20ft, taken by road or rail, en masse, to purpose-built vessels, offloaded at the receiving port and transported by lorry to the final destination. The impact on labour was obviously very significant. One giant container vessel could carry the same cargo as eight to ten conventional vessels and be offloaded in thirty-six hours at the newly built container port at Tilbury. In contrast, at the Royals it would take three weeks of a ten-hour day and ten times the number of dockers. By 1980, all of London's upriver docks had closed. Then, on 26 October 1980, the once-mighty Royal Group of Docks came to an end when the last ship discharged its cargo – the *Xingteng* from China.

23 Railways

Strictly speaking, the first railway in London was the horse-drawn Surrey Iron Railway which hauled goods from Wandsworth to Croydon. The line opened in

1803 and a specially built dock was dug at Wandsworth (MacMurray's Canal – see p. 168) where goods were transferred to or from the Thames. Due to the danger of a French attack on shipping going through the Strait of Dover, a canal was proposed to link the Croydon terminus with Portsmouth. Victory at Trafalgar removed the threat and in consequence the canal scheme was quietly forgotten. The Surrey Iron Railway was never a commercial success and all traffic ceased in 1846.

It is appropriate that the first steam-powered locomotive to travel through (or, more accurately, round) London was that of the man who invented the locomotive. Richard Trevithick was born in Camborne, Cornwall, and was described at school as 'disobedient, slow, obstinate and a loafer'. He left school barely literate and got a job in the local tin mine, where it was discovered that he could solve engineering problems, in particular those with James Watt's steam engines, many of which were used in the tin industry. The enormous size of the existing beam engines precluded their use as mobile sources of power, and Trevithick's great claim to fame was to invent the high-pressure, non-condensing engine operating at 20–25psi. It was compact and when he put it on some rails, the steam locomotive was born. In 1804, to win a bet, his locomotive hauled 10 tons of coal and seventy men for 10 miles along a tramway at Samuel Homfray's Penydaren Iron Works in South Wales. Then, in 1808, as a publicity stunt, he ran his locomotive *Catch Me Who Can* around a circular track in Euston Square. Ever the betting man, Trevithick boasted that his engine could beat any horse at the October meeting at Newmarket. And so it could, for at a minimum speed of 10mph, it could cover 240 miles in twenty-four hours. But Trevithick's railway was a circus act; the first proper railway in London ran from London Bridge to Greenwich.

London Bridge

Prompted by the success of the Liverpool & Manchester Railway, plans for the London & Greenwich Railway were authorised by Act of Parliament in 1833, the line intended as the first stage of the London to Dover route. The route ran through low-lying land prone to flooding and therefore, under the direction of the retired Royal Engineer Lieutenant Colonel G.T. Landmann, it was constructed on a massive brick viaduct of 878 arches.

It was intended to raise revenue by letting out the spaces beneath the arches. The first section of the track, between Spa Road and Deptford, opened in February 1836, and in December of that year it reached London Bridge. As was so typical of the Victorians, it was a day of great festivity. The Lord Mayor of London and City Aldermen filled the first train, as did the band of the Welsh Guards. Church bells rang and the band of the Scots Guards played the train out of London Bridge, while the party was welcomed at Deptford to the sound of music from the band of the Coldstream Guards. The journey took twelve minutes and a regular service was started. But there were issues for female passengers! Due

Aerial view of the London & Greenwich Railway. *(By permission of Southwark Local History Library)*

to the difference in height between carriage and platform, 'ladies were forced to show the beauty of their legs and ankles', much to the pleasure of conductors, who – when helping them alight – were 'apt to press the ladies' fingers and stare them full in the face'.

A railway to Croydon was built in 1839. It fed into the Greenwich line at Corbett's Lane and was followed soon afterwards by the London to Brighton route. Later, the South Eastern Railway extended the London & Croydon line to Dover. The London & Greenwich Railway cashed in on these new lines by charging a fee for every passenger carried into London Bridge over their line. Not surprisingly, the tolls were resented and so the London & Croydon and South Eastern companies diverted the line from just before Corbett's Lane to a new station at Bricklayers' Arms, close to the junction of Tower Bridge Road and the Old Kent Road. They gave the station the rather incongruous name of West End Terminus. It was not in the West End, of course, and was never a commercial success, its chief claim to fame being the place of Princess Alexandra of Denmark's arrival to marry the Prince of Wales. Both companies eventually gave up Bricklayers' Arms and negotiated a better deal with the London & Greenwich Railway to use that company's lines to London Bridge from Corbett's Lane. The latter had the distinction of having the first signalbox: 'In consequence of the trains having to pass along the Greenwich Railway ... the utmost care and vigilance are required to prevent collisions with the Greenwich trains. A Signal House is, however, placed at the junction and by judicious arrangements, accidents seldom occur.'

In 1846, the London & Croydon company experimented with a scheme of propelling their locomotive on an atmospheric railway. Pumping stations were built to create a vacuum in a pipe running between the rails. A piston within the pipe was connected to the train via a slit, sealed by a leather valve, and the pressure of the atmosphere propelled the train forward. Technical problems abounded and the scheme was eventually abandoned.

The London & Greenwich line was taken over by the South Eastern Railway in 1845, and one year later the Croydon and Brighton companies amalgamated to form the London, Brighton & South Coast Railway. London Bridge Station was rebuilt by Samuel Beazley and both companies shared it, the South Eastern to the north and – separated by a wall – the London, Brighton & South Coast to the south.

Euston

A number of sites were considered for the new London & Birmingham Railway. The first was a terminus near the Regent's Canal, enabling a link with the docks; the second, suggested by Robert Stephenson, was at Marble Arch; and the third at Maiden Lane, near King's Cross. All were rejected, making way for a terminus in Camden Town, north of the Regent's Canal, and then an extension to Euston Grove. An early problem was the incline out of Euston which necessitated, until 1844, the use of a stationary steam engine and cable to haul trains out. The grand station at Euston was planned by Robert Stephenson with Philip Hardwick as architect. But it was not to the liking of the Gothic-enamoured Augustus Welby Pugin, who wrote:

> The architects have evidently considered it an opportunity for showing off what they could do instead of carrying out what was required. Hence the colossal Grecian portico or gateway, 100 feet high … This piece of Brobdingnagian absurdity must have cost the company a sum which would have built a first rate station.

(This was the much-loved Euston Arch, demolished to the displeasure of many in 1960, when the new Euston Station was built.) Euston thus became the first mainline terminus in London, and the first to connect the capital with another major city, Birmingham. The station was rebuilt in 1846, when the London & Birmingham became part of the London & North Western Railway (LNWR). There were further extensions in 1869, when the two pavilions were added showing an alphabetic list – with much poetic licence – of towns served by the railway. In 1923, the LNWR joined with the Midland Railway to form the London, Midland & Scottish Railway (LMS).

St Pancras

The Midland Railway was based in Derby. In 1857, it had reached as far south as Hitchin and then ran over the Great Northern Railway's (GNR) line to King's Cross, paying the latter company a toll. By the 1860s, difficulties they were having with the GNR and congestion at King's Cross convinced the Midland that they should have a terminus in London. An extension was thus agreed from Bedford to a site to the west of King's Cross in the slum area of Somers Town. Work began on St Pancras Station in 1866 and took two years, during which time 10,000 people were evicted from their homes, the Fleet River was diverted through a pipe, the dead disturbed in a portion of St Pancras Church's burial ground and St Luke's Church demolished. William Barlow was the chief engineer, and the magnificent roof (689ft long and 100ft high) was by R.M. Ordish. The 240ft span of the train shed was supported by iron ribs from the firm of Butterley of Derby.

And there is also the hotel. The St Pancras Hotel is justly famous as 'the finest example of neo-Gothic architecture in London'. It is the work of Sir George Gilbert Scott, who won the architectural competition held in 1865. The well-loved building has scores of dormer windows, pointed arches and a multitude of chimneys. Scott had always wanted the commission for a government building in Whitehall. In this he was frustrated, and wrote of the Midland Hotel:

> It is often spoken of to me as the finest building in London; my own belief is that it is possibly too good for its purpose, but having been disappointed, through Lord Palmerston, of my ardent hope of carrying out my style in the Government Offices, and the subject in the meanwhile taken out of my hands by other architects, I was glad to be able to erect one building in that style in London.

King's Cross

The Great Northern Railway terminated first in 1846 at a temporary station in Maiden Lane (now York Way). The present station of King's Cross – taking its name from a statue of George IV which once stood nearby – was built on a site formerly occupied by a small pox and fever hospital. It opened in 1852. As we have seen, the Great Northern allowed the Midland to use its terminus until overcrowding and the vast increase in traffic caused by the International Exhibition of 1862 compelled the Midland to seek alternative arrangements. The talented Cubitt family built the station; Lewis as architect, and William and Joseph as engineers. The clock outside is by Dent and was previously shown at the Great Exhibition. A suburban station was opened to the west in 1875. Lewis Cubitt also designed the adjacent Great Northern Hotel.

Broad Street

In 1846, sanction was granted to build a line from the London & Birmingham line at Camden to the West India Docks and so enable goods to be conveyed from Birmingham to the docks and vice versa. It was originally called the East & West India Docks & Birmingham Junction Railway but changed its name to the North London Railway in 1853. By 1860, it had been extended to Willesden. It became very popular with City commuters who were prepared to make the long journey to the City from places in north London via Bow and then to Fenchurch Street. Realising the opportunity, the company built a spur into the City at Broad Street from Dalston Junction. The station opened in 1865 and was an immediate success – 14 million passengers used it in its first year, rising to an astonishing 46 million in 1896. But then trams and the Underground railway came, and Broad Street went into decline. The station was used by 21 million passengers in 1913, falling to 11 million in 1921. In a brave move in 1910, attempts were made to attract businessmen by opening a direct service to Birmingham and Wolverhampton, with typists on hand to take dictation. The service came to nothing and was withdrawn in 1914. Broad Street Station closed in 1986.

Liverpool Street

The Eastern Counties Railway developed in an ad hoc manner, first from Mile End to Romford and then, in 1840, a permanent terminus was opened at Shoreditch. It was renamed Bishopsgate but was not looked on particularly

Liverpool Street Station.

favourably by the Rev. Timothy Gibson who described it as 'a sort of receptacle for pick-pockets, housebreakers and prostitutes'. It was the company's wish to move closer to the centre of the City but nothing happened for many years. Eventually the go-ahead was given. Finsbury Circus was first proposed as a terminus and then Wormwood Street. In the end a station in Liverpool Street, next to Broad Street, was chosen. By this time, the Eastern Counties Railway had become the Great Eastern Railway. They built their station in 1874 on a 10-acre site, destroying in the process the gardens of the Bethlehem Hospital and demolishing the City of London Theatre, a gasworks and 450 tenements. Many people were rehoused in Homerton and Hackney, and a 2d return fare was introduced for them to come in via a junction at Bethnal Green. Ten years later, the Great Eastern Hotel opened, the work of Charles Barry.

Liverpool Street became London's largest terminus (until Victoria opened in 1908), and today handles vast numbers of commuters.

Fenchurch Street

Fenchurch Street was the first station in the City of London. It opened in 1840, when the London & Blackwall Railway was extended the short distance from its original terminus in the Minories. The railway carried passengers to the ocean-going steamers which sailed from Brunswick Pier in Blackwall to all points of the globe. It also linked the East India Dock with the City.

The early railway engines were not of the conventional variety. Instead, stationary steam engines, at each end of the line, operated ropes which pulled the passenger carriages along. At first, the rope was made of hemp and it was not uncommon for it to snap, the recoil causing quite considerable damage. As if that was not enough, the hapless passengers were then required to assist the guard by pushing the coaches to the next station! Fares were 6d for first class and 4d for second class, giving rise to the popular term for the railway as 'the fourpenny rope'.

The line opened with the usual ceremony on 4 July 1840. The first train set off from the Minories to deliver dignitaries to a grand banquet of turtle with iced punch, whitebait with champagne, and grouse with claret at the East India Company's warehouse at Brunswick Wharf. The occasion was such a sight to see that the inhabitants of two tenement houses by the line were prompted to remove their roof tiles and poke their heads through to get a better view. When conventional steam trains were introduced, a roof was erected over the Limehouse Viaduct to prevent sparks from the engine igniting the wooden ships below. The journey time from the City to Blackwall was a mere twelve minutes.

In 1849, a short extension was added from Stepney to Bow, and one year later a link was established with the East & West India Docks & Birmingham Junction Railway. Fenchurch Street then saw a vast increase in passengers coming into the City from the northern suburbs. The station was rebuilt in 1854 by George Berkeley, and in the same year the London, Tilbury & Southend Railway opened, reaching Southend in 1856. Also in 1854, the Eastern Counties Railway connected to Fenchurch Street via Bow. Thus, until Liverpool Street Station opened in 1874, Fenchurch Street was the Great Eastern's suburban line terminus.

Charing Cross

In 1859, the South Eastern Railway got permission to extend their line at London Bridge to a terminus at Charing Cross. St Thomas' Hospital lay in its path, forcing the railway company to pay £296,000 compensation which was used to build a new hospital in Lambeth on the Albert Embankment. Charing Cross Station was built on the site of the old Hungerford Market and Brunel's suspension bridge was demolished as part of the scheme. The station and new bridge were the work of John Hawkshaw. The imposing hotel was built by Edward Middleton Barry in 1865.

Cannon Street

Just as Charing Cross was the West End terminus of the South Eastern Railway, so Cannon Street was its City terminus. A spur was taken off from the Charing Cross line, and a bridge, viaduct and station were built by John Hawkshaw. The station opened in 1866, and one year later so did E.M. Barry's City Terminus Hotel.

Blackfriars & Holborn Viaduct

In 1859, the East Kent Railway changed its name to the London, Chatham & Dover Railway. The company extended its line from Herne Hill, through south London on a brick viaduct, over the Thames on Joseph Cubitt's fine bridge to Blackfriars, and thence to Holborn Viaduct and a junction with the Metropolitan Railway at Farringdon.

Victoria

Victoria Station opened in 1860, at the end of Victoria Street on the site of part of the Grosvenor Canal. It passed through the land and houses of the wealthy and there were predictable complaints. To mitigate these, the line was enclosed beneath a steel and glass tunnel from its bridge over the Thames to the station. Victoria Station was split into two halves, separated by a wall, with the Brighton line on one side and the Chatham & Dover line on the other. The station was reconstructed in 1902, opening in 1908. The year 1910 saw the spectacular funeral of Edward VII, with no fewer than one emperor and empress, seven kings, twenty princes, five archdukes and six dukes all arriving at the station with much splendour.

Waterloo

To early nineteenth-century policymakers, a canal link from Portsmouth and Southampton to London seemed an attractive proposition, avoiding the long coastal journey through the Strait of Dover. But the success of the Stockton & Darlington Railway dissuaded them from a canal link to one of rail. Accordingly, the London & Southampton Railway Co. obtained permission for a line and opened their London terminus at Nine Elms. By 1838, the line was complete to Woking. Thomas Brassey was the contractor, and Southampton was finally reached in 1840. The service proved very popular, particularly on Derby Day, when as many as 5,000 race-goers packed the trains to 'a point on the railway south of Kingston which is nearest Epsom' and walked the remainder of the way. In the late 1840s, powers were sought to build a station closer to the centre of London and, in 1848, a 2-mile extension on a viaduct of 290 arches was opened to Waterloo. The present Waterloo Station opened in 1922.

Paddington

The Great Western Railway is forever associated with the great engineer Isambard Kingdom Brunel. It was unique in its use of the 7ft broad gauge rather than the standard 4ft 8½in gauge, and this rather scuppered the initial proposal that the line should run into Euston and share a terminus with the London & Birmingham Railway, which used the narrower gauge. A temporary station was first built at Bishop's Bridge and it was into here in 1842 that Queen Victoria steamed on her first journey by train from Slough. Accompanying her was Brunel himself. Paddington Station was eventually designed by Brunel and Matthew Digby Wyatt and opened in 1854. At this time, the Great Western still retained its broad-gauge track, but in 1871 a third rail was inserted between the wider broad gauge so that it could take trains from other companies. This process continued until 1892, by which time the entire Great Western rail network ran on the by-now standard gauge.

Marylebone

The Great Central Railway from Manchester originally fell short of London and ended at its first terminus at Canfield Place, near Finchley Road. The 2-mile extension to Marylebone was bitterly opposed by the rich of St John's Wood and the MCC at Lord's, but, despite protests, it eventually opened in 1899.

GER Railway Works, Stratford. *(By permission of Newham Local Studies Library)*

Stratford Railway Engineering Works

Railway locomotives were made for the Eastern Counties Railway Co. at their Stratford Works and by 1847, 1,000 workers were employed. The Great Eastern Railway (GER) absorbed the Eastern Counties Railway Co. and by 1906, 6,000 men were employed on a 78-acre site. In 1891, the works set an all-time record when it built a Class Y14 engine in nine hours forty-seven minutes. The locomotive went on to give sterling service on the Peterborough to London line for forty years; superintendent at the time was James Holden. The Stratford depot closed in 2001 to enable work to be undertaken on the cross-Channel railway link. The land now forms part of the London 2012 Olympic and Paralympic Games site.

London Underground

Charles Pearson was born in 1793, the son of an upholsterer and feather merchant. He qualified as a solicitor in 1816 and spent much of his life campaigning. He fought against corrupt jury practice, for the right of Jews to become bankers, against capital punishment, for the disestablishment of the Church of England, for universal suffrage and much more. Pearson also wished 'to relieve the congestion of London's streets'. By the mid-nineteenth century hundreds of thousands of people were pouring into London every day, many of them by rail. His plan was for an underground railway to link the Great Western's terminus at Paddington with the City at Farringdon via the London & North Western at Euston and the Great Northern at King's Cross.

The North Metropolitan Railway Co. was duly formed. The first plan was to propel carriages through the tunnel by compressed air; the second – to avoid noxious smoke – was to charge locomotives at each end of the track before they travelled without a firebox through the tunnel. Both schemes came to nothing. The inevitable was finally accepted and steam trains eventually ran through a cut-and-cover tunnel. (The cut-and-cover technique involved digging a trench just beneath the surface, laying the track and then putting a roof over the top and covering the whole thing in. The spoil was used to build the terraces at Chelsea's Stamford Bridge football ground.) Despite its nickname – 'the sewer railway' – the venture was a great success and almost 1 million passengers were carried in the railway's first year, 1863. Regrettably, Pearson did not live to see the completion of his dream; he died in 1862.

In 1864, a Joint Parliamentary Committee recommended that an 'Inner Circle' be built to serve all railway termini north of the river. Accordingly, the Metropolitan District Railway was formed and built the first section of the District Line from South Kensington to Westminster. It was later extended to Blackfriars at the same time as the Thames Embankment was built. Meanwhile, the North Metropolitan had extended west to Hammersmith and east to Moorgate. The Circle Line was finally completed in 1884, and expansion in other directions continued apace, to Ealing, Richmond and Wimbledon.

Electricity came to London in the late nineteenth century, prompting the construction of the world's first Underground 'Tube' railway. The City & South London Railway ran from King William Street in the City to Stockwell through two tunnels built using James Henry Greathead's tunnelling shield. Cable haulage was proposed but was rejected in favour of electric traction via a third rail beneath the train. It was a somewhat claustrophobic journey – there were narrow windows, high up, leading to the carriages being called 'padded cells'. Nevertheless, the line was an unqualified success, attracting 5 million passengers in the first year of operation. Plans were soon made to extend the line. King William Street Station closed and a new station was built at Bank (receiving trains in a new tunnel which ran via London Bridge). By 1901, the line extended south to Clapham Common and north to Moorgate, Old Street, City Road (now closed), Angel and later (1907) to King's Cross. In time, it became the Northern Line.

London's next Tube line was the Waterloo & City Railway, built to transfer commuters from Waterloo mainline station to Bank and the City. It opened in 1898 and was owned by the London & South West Railway Company.

Next came the Central London Railway, otherwise known as the 'Twopenny Tube'. The intention was to build a tunnel from Queensway to Bank. Despite vociferous objections from St Paul's Cathedral, worried about the foundations, and Sir Joseph Bazalgette, concerned about his sewage system, the scheme – under the direction of James Henry Greathead – went ahead. The line was extended to Wood Lane in 1908 to serve London's first Olympic Games at White City

London Transport's Lots Road Power Station.

Stadium, and in 1912 to Liverpool Street. In common with many Underground lines, it was absorbed into the Underground Electric Railways Co. of London in 1917, and taken into public ownership in 1933.

In 1900, Charles Tyson Yerkes – stock market speculator, convicted larcenist, blackmailer and former prison inmate – arrived in London. By fair means or foul, he had made his money by building Chicago's transport system. Yerkes gazed down on London from the heights of Hampstead Heath and determined to get involved with London's expanding Underground railways. His first purchase was the Charing Cross, Euston & Hampstead Railway Co. which had recently been granted permission to build a deep Tube line from Charing Cross to Hampstead and Highgate. Within a year, Yerkes set his sights on the Metropolitan District Railway, proposed its electrification and renamed it the Metropolitan & District Electric Traction Co. Next came the Brompton & Piccadilly Circus Railway Co. which had plans to run from South Kensington to Piccadilly Circus. This latter company also acquired the Great Northern & Strand Railway and with it the right to build a Tube line from the Strand to Finsbury Park. The two companies' lines were eventually linked and were the forerunner of today's Piccadilly Line. However, it did not end there. In 1902, Yerkes completed his buying spree by rescuing the Baker Street & Waterloo Railway, planned to run from Paddington to Elephant & Castle, which had got into financial difficulty. Yerkes brought all these companies together, using what has been termed 'novel financial instruments', and so was born the Underground Electric Railway Co. of London (UERL).

Power for the electrified Underground was provided by Fulham's Lots Road Power Station, still standing today on the river bank. It was built by UERL between 1902 and 1904, was coal fired and had an eventual capacity of 50MW. In the 1960s, coal firing was replaced by fuel oil, probably a costly mistake because of later oil price rises. The station went out of service in 2002, and now all power for the Underground system is taken from the National Grid.

Despite financial problems in the early twentieth century, construction of the Tube proceeded apace and the outer suburbs were soon reached. There were distinctive stations as well; for example, those in the centre of London designed by UERL's architect Leslie Green, which are faced with attractive red glazed terracotta. Holloway Road, Caledonian Road and Russell Square are good examples. Charles Holden designed many of the stations in outer London and was also responsible for the commanding new headquarters at 55 Broadway, above St James's Park Station.

In 1933, UERL (and the Metropolitan Railway) were taken into public ownership, with Albert Stanley as chairman and Frank Pick as chief executive. The new organisation was known as the London Passenger Transport Board (LPTB). It was at about this time that Harry Beck designed the now familiar and, as it turned out, iconic Underground Map. After the Second World War, the LPTB (or London Transport, as it was usually known) was nationalised.

The latter half of the twentieth century saw further expansion. The Victoria Line was opened in the late 1960s, the Piccadilly Line was extended to Heathrow, and the Jubilee Line was extended through Docklands to Stratford. There were many further reorganisations, and now the Underground is operated by a Public-Private Partnership. The infrastructure and trains are maintained by Metronet and Tube Lines under long-term contracts, while the Underground is publicly owned by Transport for London.

Part 4

OTHER INDUSTRIES

24 Brewing & Distilling

Brewing is an ancient art. It was understood by the Mesopotamians who were practising malting and fermentation by 3000 BC. And although wine production predominated in the Roman Empire, brewing was known to the peoples of northern Europe, as we learn from the Roman historian Pliny:

> The nations of the west have their own intoxicant from grain soaked in water; there are many ways of making it in Gaul and Spain and under different names though the principle is the same. The Spanish have taught us that these liquors keep well.

The Roman Emperor Julian sampled British ale on a visit to Paris. He did not appreciate it and wrote a poem about his experience, 'On Wine Made from Barley':

> Who made you and from what?
> By the true Bacchus I know you not.
> He smells of nectar
> But you smell of goat.

Beer today is still brewed according to the same ancient principles. The only difference is that there is now a degree of automation and, following their introduction from Flanders, hops are added to give beer its bitter taste and to act as a preservative. In brief, barley grains are steeped in water to begin the process of germination, releasing enzymes which aid in the conversion of starch in the barley to sugar. The barley is then dried and gently crushed to expose its inside. The malted barley, now known as grist, is heated with hot water to continue the conversion of starch to sugar. A clear, sugary liquid is formed called wort, which is filtered off. The wort is then boiled in large vessels, known as coppers, and hops are added. Before fermentation the wort has to be cooled by passing it through

counterflow heat exchangers. The cooled wort is then transferred to fermentation vessels and yeast added to ferment the sugar to alcohol. Fermentation time
and temperature vary according to the type of beer being brewed, ale being fer-
mented at a higher temperature than lager. All that remains is for the beer to be
transferred to casks, kegs, bottles or cans.

Beer was consumed by everyone in the Middle Ages, mainly because water – in
London typically taken from the Thames – was infected, and tea and coffee were
unheard of. The brewing process produced a drink free of bacteria and pathogens.
Hops were introduced to England in the early 1500s and at the same time there
was an influx of Flemish hop growers into Kent. Traditional British ale is a heavy
and sweet liquor and the introduction of hops was not universally appreciated, as
John Taylor, an ale house keeper in Phoenix Alley, off Long Acre, explains: 'Beare
is a Dutch boorish liquor, a thing not known in England, till of late days an alien
to our nation, till such times as hops and heresies came about us; it is a saucy
intruder in the land.'

It was the usual practice – and certainly before the addition of hops became
the norm – for ale to be produced on the spot by inn and ale house keepers to
be consumed locally or on the premises. In 1700, half of the beer in the country was produced in this way. But by 1850, the situation had changed and the
amount brewed locally fell by 80 per cent. The reason was the emergence of
common brewers who were able to exploit economies of scale as well as intro-
duce technical advances; for instance, thermometers, hydrometers and steam-
driven machinery. They also began to tie their businesses to retail outlets, either
by purchasing public houses en bloc or lending money to publicans in return for
taking their beer.

In contrast to gin, which in the mid-eighteenth century was blamed for the
deaths of countless children – Hogarth's 'Gin Lane' cartoon of 1751 illustrates
the point – beer was seen as a wholesome drink. To quote the Prince Regent:
'Beer and beef has made us what we are.' The consumption of beer was given
another boost in 1830, when, as a device to curb the consumption of spirits and to
discourage the riotous behaviour associated with the delays in passing the Great
Reform Act of 1832, the government abolished beer duty and decreed that any
ratepayer who paid 2 guineas excise licence could sell beer. The result was a vast
increase in beer sales.

In the early nineteenth century, London brewers dominated, but their position began to be challenged by beer from Burton upon Trent in Staffordshire.
Beer had been brewed in Burton for centuries, its breweries being blessed with
water high in calcium sulphate, which gave a bright, strong and distinctive ale.
Furthermore, when the railways came in the 1840s, Burton brewers were able to
export their beer throughout the country.

The consumption of beer reached a peak around 1870, when 'going to the pub'
was the traditional, and probably only, working-class pastime. Thereafter, demand
fell, with competition coming from football, music halls, railway excursions and,
of course, pressure from temperance societies. The number of licensed premises

A London ale house – 'The Pitt Head', Bankside.

fell dramatically by the turn of the century, mainly because licences were removed by magistrates. So alarmed were the brewers that they pressed for, and got, an Act of Parliament to provide a compensation scheme.

Further pressure came in the First World War when Lloyd George blamed the brewers for a shortage of munitions, exclaiming that 'the lure of drink was doing more damage in the war than all the German submarines put together'. Opening hours were slashed and duty increased. The figures are stark: in 1919, at a national level, 13 million barrels of beer were produced, compared with 36 million in 1913. There were now many alternatives to the pub. In 1930, 40 per cent of the population went to the cinema every week, and radio encouraged people to stay at home. Britain was 'a temperate nation', said Lord Woolton, Minister of Food, in 1940.

After the Second World War amalgamations of breweries became the norm and towards the end of the twentieth century, lager beers became much more prevalent, taking 40 per cent of the market by 1980.

Well-Known London Brewers

Truman, Hanbury, Buxton & Co.
Today, Brick Lane is famous for its Bengali restaurants but in the seventeenth century it was very different. Daniel Defoe described it as 'a deep dirty road frequented chiefly by carts fetching bricks into Whitechapel'. It was here in Black Eagle Street – a passage running off Brick Lane – that Joseph Truman opened his brewery in 1679. The firm was expanded by his son, Benjamin, and by 1760, Truman's was London's third largest porter brewer, narrowly behind John Calvert's Peacock Brewhouse and the brewery of Samuel Whitbread.

Knighted by George III, Sir Benjamin Truman lived the life of a country gentleman at his estate at Pope's Manor in Hertfordshire, which was only a stone's throw from Bedwell Park, Essendon, home of his great rival, Samuel Whitbread.

In 1789, Samuel Hanbury joined the firm, living in the brewer's house in Brick Lane. Towards the end of the eighteenth century, the brewery's fortunes declined but were revived when Hanbury's nephew, Thomas Fowell Buxton, became a partner in 1811. Under his stewardship sales soared and by 1827, Truman's had overtaken Whitbread's, selling 211,000 barrels every year. (They were still way behind Barclay Perkins, however, which, at that time, were the largest brewers in the world.) Thomas Fowell Buxton was an eminent man. He became a baronet, was MP for Weymouth, worked for the abolition of the slave trade, campaigned for prison reform and succeeded William Wilberforce as head of the anti-slavery party in 1824. He is commemorated by a Gothic fountain in Victoria Tower Gardens in Westminster. As so often happens, his son was a complete contrast and was described thus: 'As he is one of the richest men in the country, so he is the meanest. So thoroughly is he despised ... not a village boy touches his hat when the wealthy brewer passes.'

Truman, Hanbury & Buxton, Brick Lane Brewery, 1842. *(By permission of Tower Hamlets Local History Library)*

In response to the threat posed by beer brewed at Burton upon Trent, Truman's purchased Phillips Brewery in Burton in 1873, and throughout the twentieth century, in common with all large brewers, took over many smaller concerns.

In 1970, Truman's contracted with the Danish brewer Tuborg to brew their lager at Brick Lane and to carry out a vast modernisation programme, rebuilding much of the Brick Lane premises. It was at this time that Truman's became the centre of what was described in *The Observer* newspaper as 'the most incredible takeover battle of all time'. The contestants were the Grand Metropolitan Hotel chain and the West London brewers Watneys. The battle even prompted a sermon from Canon John Collins at St Paul's Cathedral. Both parties were criticised and Canon Collins pointed out that both had little consideration for the national interest, let alone the interests of Truman's workers or its small shareholders. In the end, Grand Metropolitan won the day and took control of Truman's. New offices were built under the direction of Ove Arup in 1980; fine as they were, beer sales still fell. The Brick Lane Brewery eventually closed in 1989.

Watney

In 1722, a Mr Green of Pimlico Lodge built himself 'the finest brewhouse in Europe', and perhaps because he had three stags incorporated in his coat of arms, he named his brewhouse the Stag Brewery. John Elliot joined in 1795, to be followed in 1837 by James Watney, who eventually bought out Elliot, the firm now going under the name of Watney & Co. In the late nineteenth century, Watney took over the smaller breweries of More & Co. of Old Street, and Carter Wood & Co.'s Artillery Brewery in Westminster. But in terms of its later development, the most important acquisition was the Mortlake Brewery of Phillips & Co., which was said to date from 1487. Then, in 1898, Watney's combined forces with Reid & Co. and Combe & Co.

The Reid family were originally in partnership with the Meux family at the Griffin Brewery in Liquorpond Street, off Clerkenwell Road. A violent row between the two families in 1809 split the partnership and the Meuxs left to set up at the Horseshoe Brewery in Tottenham Court Road.

Oswald Serecold joined the firm in 1886. He tells us of the strict dress code enforced by Reid's. Directors, office workers and brewers all wore top hats. If they turned up with any other headgear they were fined 1 guinea. Draymen had characteristic red caps and white top coats. The firm's brewing techniques were shrouded in secrecy to the point of paranoia – thermometers were calibrated with letters rather than degrees – and the same level of secrecy extended to the firm's accounts. No one, not even the shareholders, saw them apart from the directors. Reid's brewery eventually closed and the site was sold to the London County Council which built houses on the site.

Combe & Co. of Castle Street, Long Acre, began as the Wood Yard Brewery, founded in 1740 by a timber merchant. Harvey Christian Combe, with others, bought the brewery in 1787 and it became known as the Gyfford's Brewhouse. Harvey Combe (later to become Lord Mayor of London) loved entertaining and, in 1807, he treated the Prince of Wales and other members of the royal family to a beefsteak dinner, washed down, no doubt, with his famous porter. He was also a gambling man, as was the great radical parliamentarian Charles James Fox. On one occasion, Fox 'won a fairly considerable sum' from Combe after which he retorted, 'Bravo, Old Mashtubs, I will never drink any other porter than yours.' To which Combe responded acidly, 'Sir, I wish every other scoundrel in London would say the same.'

All three breweries combined in 1898 to form what was then London's largest brewery, Watney Combe Reid & Co. At the time of amalgamation, beer was delivered by teams of dray horses; 260 horses were stabled at the Stag Brewery and 40 at Mortlake. A new malt house was soon built at Mortlake with the capacity to produce 10,000 quarts of malt per year.

Early in the twentieth century, Watney's acquired several smaller London breweries, such as the Welch Ale Brewery in Chelsea, Huggins of Soho, and Stepney's London & Burton Brewery. In 1930, they introduced one of their best-known brews, launching a competition in the evening papers to find a name for it. A £500 prize was offered, and 26,000 hopeful punters tried their luck. The winning entry came up with the name 'Red Barrel' – a plain barrel, printed in red, with the word 'WATNEYS' written across it in block letters. Later in the same decade, Watney's began an advertising campaign with a series of eye-catching and ingenious 'brick wall' posters. The 'wall' would have a chalk message written on it; for instance, 'Stick no bills, stick to Watneys' and 'What we want is Watneys'.

In 1958, Watney's acquired the Albion Brewery of Mann, Crossman & Paulin in Whitechapel Road and became known as Watney Mann Ltd. The company was bought by Grand Metropolitan Hotels in 1972 and two years later merged with Truman, Hanbury, Buxton & Co.

Courage's

The Revocation of the Edict of Nantes in 1683 accelerated the persecution of Protestants already rife in France. Many sought refuge in Britain, among them the Courage family, who settled in Aberdeen. In 1780, John Courage, son of Archibald, arrived in London as agent for the Glasgow-based Carron Shipping. It was not long before he paid John and Hagger Ellis £616 13s 11d for their river-side brewhouse in Horsleydown, Bermondsey.

Courage's drew water from a well 450ft in depth. This was fine for mild ale and stout, but competition was increasing from the pale ales brewed at Burton. London water lacked the appropriate minerals to make pale ale, so in 1872, Courage's came to an arrangement with Flower's of Stratford upon Avon for them to supply pale ale. Later, they came to a similar arrangement with Fremlin's of Maidstone. Then, rather than purchase a brewery in Burton (as other London brewers had), Courage's bought the G. & E. Hall Brewery in Alton to make pale ale.

Courage's Brewery, Horsleydown, Bermondsey.

Disaster struck at the Horsleydown brewery in 1891 when a spark in the malt house caused an explosion and fire. For the next few weeks, until the brewery was rebuilt, Courage's had to take beer from their nearby rival Barclay Perkins (and pay them £40,000 for it).

In common with other large brewers, Courage's absorbed smaller concerns in the early twentieth century. In 1955, they merged with Barclay Perkins, took over Simonds of Reading in 1960, and John Smith of Tadcaster in 1970. They were acquired by the Imperial Tobacco Group in 1972. Brewing stopped at Bermondsey in 1982.

Whitbread

Samuel Whitbread was born in 1720 in Cardington, Bedfordshire. Educated by a local clergyman, he came to London at the tender age of 14 and was apprenticed at John Lightman's Brewery in Clerkenwell. Always ambitious, he went into partnership with Godfrey and Thomas Shewell and opened two breweries off Old Street. The Goat Brewery was on the corner of Whitecross Street and nearby was a small brewery in what is now Central Street (then Brick Lane). Within a few years, Samuel Whitbread noticed a disused brewery in Chiswell Street. He bought it and promptly built a new brewery for the mass production of porter. Regulations to curb the consumption of gin boosted sales of porter – beer was considered a healthy drink – and by 1758, Whitbread's was the largest brewer in London.

Samuel Whitbread was a man of high morals. He read the scriptures every day, observed the Sabbath, supported the anti-slavery movement and the work of prison reformer John Howard. It was always his intention for his son, also Samuel, to take over the firm. He wrote to his son:

> Your father has raised it, from a very small beginning and by great assiduity in a very large course of years … and with the highest credit in every view by honest and fair dealing … and the beer universally approved. There was never the like before, nor probably ever will be again, in the Brewing Trade.

But Samuel Whitbread II had no business sense or interest in brewing. Whitbread's went downhill fast – only 100,000 barrels were produced in 1809, compared with 202,000 under Whitbread senior. Samuel's mental condition went from bad to worse and he eventually cut his own throat in 1815. The business was saved, however, by the fortunate amalgamation with the well-established brewery of John Martineau of Lambeth. Martineau turned things around and by 1823, Whitbread was back and producing 213,000 barrels. Poor Martineau was to come to an unfortunate end, however. He fell into a vat of beer and was pronounced dead 'by the visitation of God'.

Whitbread's became a private limited company in 1889, and a public limited company in 1948. Brewing ceased at Chiswell Street in 1976.

Charrington

The story begins with Robert Westfield and Joseph Moss, who were running a brewhouse in the Mile End Road. Meanwhile, John Charrington, son of a Hertfordshire vicar, was learning his trade at a brewery by the name of Hale in Islington. Then, in 1776, John Charrington, for the considerable sum of £1,446 4s 4d, acquired a one-third share in the Mile End Brewery. By 1783, with his brother Nicholas, he was in full control.

The brewery became known as the Anchor Brewery and in common with other breweries began a process of expansion, absorbing other breweries along the way, such as Steward & Head of Stratford upon Avon. By the mid-nineteenth century, Charrington's had achieved a thirtyfold expansion.

But there was a rebel in the family. Frederick Nicholas Charrington became a temperance evangelist. The turning point was when he observed a raggedly dressed woman with her children in tow call her husband out of a pub and ask for money to feed her starving children. The brutish husband responded by knocking her into the gutter, at which point Frederick looked up and saw the name Charrington writ large on the front of the pub. He vowed never again to enter the brewery and began a crusade against the demon drink, prostitution and music hall entertainment. In the event, his brother bought him out and, ironically, Frederick used the money, presumably accumulated through the sales of beer, to finance his temperance work.

In 1933, Charrington's took over Hoare's, one of London's oldest breweries. Their history can be traced back to 1492, when Henry VII licensed the Fleming

Charrington's Brewery, Mile End Road. *(By permission of Tower Hamlets Local History Library)*

John Merchant to export 50 tuns of ale. The firm operated from the Red Lion Brewery in Lower East Smithfield. It was later acquired by Sir John Parsons, Lord Mayor of London in 1703 and MP for Reigate. Their beer was a firm favourite with Oliver Goldsmith who called it 'Parson's black champagne'. The Hoare family entered the firm in 1802, when Henry Hoare, a partner in Hoare's Bank, bought a share for his son. In 1923, Hoare's took over the long-established Lion Brewery of Lambeth, which ceased brewing at the same time. It is interesting to note that the Coade stone lion which stands at the eastern end of Westminster Bridge once stood proudly, painted red, in front of their brewery.

The sale of Hoare's to Charrington's took place with the minimum of fuss. It was a cold winter's day. The Whaddon Chase hounds were drawing a covert. Guy Charrington, meanwhile, was sitting patiently on his horse beneath a tree. Up rode William Murray, a director of Hoare's. With few words the deal was done.

Brewing ceased at the Red Lion Brewery in 1934, and at the Anchor Brewery in 1975.

Barclay Perkins

Throughout its history, Bankside has been associated with the brewing industry. The miller summed it up in Chaucer's *Canterbury Tales*:

And if the words get muddled in my tale,
Just put it down to too much Southwark ale.

Later, in the seventeenth century, dramatist Thomas Dekker wrote of Bankside: 'The whole street is a continual ale house.'

Bankside's most famous brewery was Barclay Perkins. The firm's history goes back to the mid-seventeenth century when James Monger, 'Citizen and Clothworker of London', opened a brewhouse near the site of the Globe Theatre. Ownership then passed to James Child, a liveryman of the Grocers' Company, who, in 1670, practised 'the art and mistery of brewing'. Then, towards the end of the century, an enterprising young man took a job as a broom clerk, involving no more than sweeping up, but within two years Edmund Halsey had married the boss' daughter and in 1693 took over the business.

Halsey was a man of many parts. He was certainly an astute businessman, for the brewery thrived under his direction. He also had political ambitions, but in 1710 was accused of bribery and was thus prevented from taking his seat as MP for Southwark. In 1719, he was governor of St Thomas' Hospital and was elected MP for Southwark in 1722 and again in 1727. Halsey had no son to carry on the business and so his nephew, Ralph Thrale, was brought in, who, in the words of Dr Samuel Johnson, 'was employed for 30 years at 6s a week in the brewhouse that was afterwards his own'. And he paid for it as well, for Halsey's will directed that 'all my stock in trade shall be sold for the best price that can reasonably be gotten'. The price was £30,000 – an indication of the brewery's growth, for it had changed hands for £400 a hundred years before.

Ralph Thrale was an enterprising man, MP for Southwark, High Sheriff of Surrey and Master of the Brewers' Company. He amassed a fortune, enabling him to educate his son, Henry, at Oxford, of whom Johnson remarked that 'although in affluent circumstances, [he] had the good sense enough to carry on his father's trade'. Henry Thrale inherited the business in 1758 and was soon to marry a remarkable lady, Hester Lynch Salusbury – talented, quick-witted and with a lively, artistic mind much interested in the arts. Dr Johnson was a firm friend of the family. In 1773, a special beefsteak dinner was served at the brewery in the company of Johnson, Sir Joshua Reynolds, Oliver Goldsmith, David Garrick and Edmund Burke. Mrs Thrale was enthralled by Johnson and his circle. Johnson had his own room set aside at the brewhouse and even advised on its running. During a period of financial difficulty he wrote to Mrs Thrale: 'The first consequence of our late trouble ought to be an endeavour to brew at a cheaper cost ... Unless this can be done nothing can help us.'

Meanwhile, the day-to-day running of the brewery was left to its long-suffering manager, John Perkins. Perkins' moment of glory came in June 1780. While the Thrales were away taking the waters in Bath, the brewery became the target for mob violence. Thrale, following in the family tradition as MP for Southwark, had shown sympathy for Roman Catholic emancipation and this motivated the Gordon Rioters to seek reprisal. Fresh from their assault on Newgate Prison, the mob surged south of the river. There they were met at the brewery gate by Perkins who, according to Boswell, bought them off with 'fifty pounds worth of meat and porter'. They returned, of course, but by this time Perkins had called in the troops. He was hero of the hour!

Henry Thrale was soon to die and Hester determined, with Johnson's help, to sell up. 'For Sale' signs were stuck on the brewery walls and these were noticed by David Barclay (of Barclay's Bank fame) while out walking. 'This will do for young Robert,' he exclaimed – Robert being his American-born nephew. The day of the sale was fixed for 31 May 1781. Johnson, as executor to Thrale's will, was, according to Lord Lucan:

> ... bustling about, with an ink horn and pen in his button hole, like an excise man, and on being asked what he really considered to be the value of the property answered 'We are not here to sell a parcel of boilers and vats but the potentiality of growing rich beyond the dreams of avarice.' Mrs Thrale was well pleased to get £135,000, exclaiming 'God Almighty sent us a knot of rich Quakers who bought the whole.'

The Barclays were astute enough to take John Perkins in with them with a quarter share, and the brewery became known as the Anchor Brewery, changing to Barclay Perkins & Co. in 1791.

Around this time, the 'horse wheel' that raised water from the brewery's well was replaced by a Boulton & Watt steam engine. This was installed by William Murdoch, generally recognised as the founder of the gas industry, but at the time working for

Boulton and Watt. Barclay wanted some indication as to how the work done by the new steam engine compared with the horses he was using at present. Murdoch calculated one horsepower as 33,000ft-lb/min, an engineering definition born in Bankside that was to last until the days of metrication!

Barclay Perkins went from strength to strength. In the nineteenth century there were many famous and indeed infamous visitors. One such was the brutal Austrian general Julius von Haynau, notorious for his flogging of women. The draymen did not take kindly to his reputation, chasing him off the premises. The mob joined the hunt, forcing the general to take refuge in a dustbin. There were demands for a formal apology but Foreign Secretary Lord Palmerston took the draymen's side, much to the dismay of Queen Victoria. The incident so thrilled the Italian patriot Garibaldi that, on his visit to London in 1864, he asked to see the brewery where 'the men flogged Haynau'. A plaque in Park Street, Bankside, marks the spot today.

In the twentieth century there was further expansion and other breweries were taken over, such as the Royal Brewery at Brentford, Style & Winch, and the Dartford Brewery Co. In 1955, Barclay Perkins merged with Courage and finally closed in 1982, the site being purchased by the GLC for £2.5 million, bringing to an end over 250 years of brewing in Bankside.

Mann, Crossman & Paulin

James Mann was born in 1776 and entered the brewing industry as a partner of Philip Betts Blake at the Stanbridge Brewery in Lambeth. Meanwhile, in 1808, next to the 'Blind Beggar' ale house in Mile End, Richard Ivory built a brewery which he called the Albion Brewery. He leased it to John Hoffman. For whatever reason, Hoffman did not make a success of the job, lost money and eventually ended up bankrupt, and the Albion Brewery came up for sale. Philip Betts Blake (with James Mann) stepped in and his bid of £2,420 was accepted. Blake later sold his share to Mann who was, in due course, succeeded by his son James junior.

Robert Crossman and Thomas Paulin were soon to join the Albion Brewery, and the firm became known as Mann, Crossman & Paulin. Both Crossman and Paulin were from Berwick in Northumberland; they were school friends and both learned their trade at a brewery in Isleworth, Middlesex. Until he joined Mann in Mile End, Paulin remained at Isleworth, but Crossman returned to Berwick as manager of the Border Brewery in that town.

As was the case with all their competitors, Mann, Crossman & Paulin began to accumulate tied houses in the mid-nineteenth century, including the famous 'King's Head' in Chigwell, immortalised as 'The Maypole' in Charles Dickens' *Barnaby Rudge*. While still remaining next door to the 'Blind Beggar', the firm began a process of expansion to include the installation of a device called 'Juke's Patent for Consuming Smoke'. Evidently the brewery had been getting complaints for polluting either the beer or the neighbourhood, but it seems the contraption (whatever it was) was ineffective as 'this had to be taken down not answering the purpose for which it was erected'.

In 1875, Mann, Crossman & Paulin purchased 30 acres of land in Burton upon Trent, and in common with other London brewers, opened in that town with a brewery also known as the Albion Brewery. And it was from their Burton branch that Thomas Thorpe was recruited to come to London as manager and revitalise the Mile End brewery. Thorpe was dynamic and dictatorial, and began a policy of acquiring large numbers of beerhouses and off licences in London's back streets. At a similar time, the Albion Brewery became the first in London to brew Brown Ale.

On 2 January 1911, the Albion Brewery and those who worked there were caught up in the siege of Sidney Street. A couple of weeks before, in what became known as the Houndsditch Murders, three members of the Metropolitan Police were shot dead by an armed gang of Latvian revolutionaries. It was rumoured that two of the gang, one of whom was said to be Peter Piaktow (otherwise known as Peter the Painter), were holed up at 100 Sidney Street next to the brewery's bottling store. The police made no attempt to arrest the men, but with the building surrounded by a force of 200 and from the cover of the brewery yard, they threw pebbles up at the window to induce the men to leave. Rather than quit the building quietly, the response from the anarchists was a volley of shots and the siege of Sidney Street began. It raged for six hours and Winston Churchill, who had arrived at the scene to witness it for himself, called in the Scots Guards. In the end, the building caught fire and the two anarchists perished. Neither of the victims was Peter the Painter, who in due course turned up in the Soviet Union as deputy head of the Cheka, the Soviet state security organisation. Later, he perished in the Great Purge of 1938. As for the brewery, *The Times* newspaper reported, 'an innocent drayman putting his horses in the yard narrowly escaped a bullet, a wounded police sergeant could only be conveyed to hospital by passing a stretcher, which the brewery fortunately supplied, over the back wall'.

Mann, Crossman & Paulin also purchased a series of smaller breweries in the early twentieth century. In 1920, they took over Brandon's Brewery of Putney, together with its branch at Twickenham, and put Francis Thomas Mann (famous as a test match cricketer) in charge. Best's Brewery of Camberwell and The Brewery at Southend-on-Sea were absorbed later in the decade. In 1958, the company merged with Watney Combe Reid to form Watney Mann Ltd. Brewing eventually ceased at the Albion Brewery in 1979.

Young & Co.

The first mention of brewing at Wandsworth was in 1581 when Humphrey Langridge is recorded as brewing beer at the sign of 'The Ram'. In 1670, the Draper brothers, Somerset and Humphrey, took over, and in their reign at Wandsworth the fine eighteenth-century brewhouse was built. It still stands today. One hundred years later the brewery was in the hands of the Tritton family, who began a process of expansion and started to brew porter. Beer was delivered from Wandsworth by the traditional horse and dray. Then, in 1803, the pioneering Surrey Iron Railway opened from Wandsworth to Croydon and this provided an additional means of distribution.

In 1831, the Young family entered the business when Charles Allen Young and Anthony Fothergill Bainbridge bought out the Trittons. One year later, the brewery suffered a disastrous fire, but rebuilding was swift and by 1835 was complete, with a new beam engine installed as well. It is still in good working order and continued to raise steam until 1976. Young and Bainbridge dissolved their partnership in 1883 – hardly surprising since Bainbridge had run off with Charles Young's wife. Young remained, however, and the firm continued to flourish throughout the twentieth century, winning many awards for its fine 'real ale'. By the beginning of the twenty-first century, the Wandsworth site was becoming ever more difficult to operate and, in 2007, Young's decided to sell and team up with the Bedford brewery of Charles Wells.

Fuller, Smith & Turner

Beer was brewed in the gardens of Bedford House, Chiswick Mall, in the seventeenth century. Later, Douglas and Henry Thompson and Philip Wood set up a brewery in Chiswick, but by the early nineteenth century got into financial difficulty. They approached John Fuller to help them out, who, after making a money donation, entered into business with Douglas Thompson. Thompson, however, soon fled the country, leaving Fuller with the job of finding a partner with brewing experience. Henry Smith and his brother-in-law John Turner fitted the bill. They both came from the Romford Brewery of Ind & Smith and so it was that Fuller, Smith & Turner was founded in 1845.

In the early twentieth century, the Beehive Brewery in Brentford was taken over and 100 years later George Gale & Co. of Horndean in Hampshire were

Fuller's Brewery, Chiswick, in the 1960s.

acquired. Now the only brewers in London, Fuller's have won many prizes for their beers, London Pride, ESB and Chiswick Bitter. In 1995, they published their history in a book titled *London Pride*.

Other London Breweries

A brewery was founded in the early eighteenth century in St John Street, Clerkenwell, and by the middle of the century took the name Cannon Brewery. It changed hands several times and in 1930 was taken over by Taylor Walker, when brewing ceased in St John Street.

Taylor Walker brewed at Church Row, Limehouse, until 1960. The firm began as Stepney Brewery in 1730 and became Taylor Walker in 1816. In 1823, operations were relocated to Fore Street in Limehouse and then, in 1889, moved to the Barley Mow Brewery in Church Row, Limehouse. Taylor Walker were acquired by Ind Coope in 1960.

As we noted earlier, Henry Meux was in partnership at the Griffin Brewery of Reid, Meux & Co. in Liquorpond Street, Clerkenwell. He left in 1809, following a dispute with the Reid family, and opened up at the Horseshoe Brewery in Tottenham Court Road on the corner with Oxford Street. It brewed porter in a vast vat, 22ft high and with a capacity of 3,555 barrels – enough beer to give 1 million Londoners one pint of beer each. In 1814, it burst with disastrous consequences, spilling its contents into adjacent houses. The area contained some of the poorest slums in London. Several rundown tenements were flooded and eight people drowned. The beer flowed out into Tottenham Court Road and hoards of Londoners descended on the scene, scooping the beer up to drink and getting intoxicated in the process. But Meux's were less than happy with the accident – they had already paid the government duty on the lost beer. Not wishing to lose such a large sum of money, Henry Meux applied successfully to Parliament to have the duty reimbursed. In 1914, Meux's acquired the Nine Elms brewery of Thorne Brothers and transferred all brewing there in 1921, the Horseshoe Brewery in Tottenham Court Road closing at the same time. Later, in 1956, the company joined with Friary, Holroyd & Healy's brewery of Guildford and became Friary Meux Ltd. Brewing came to an end at Nine Elms in 1964.

The City of London Brewery, once one of London's largest, was situated at 89 Upper Thames Street, at the Hour Glass Brewery. It was in existence as long ago as 1431, and in 1587 was run by John and William Reynolds. Stow tells us: 'At the east end of the church of All Hallows the More, goeth down a lane called Hay Wharfe, now lately a great brewhouse builded there by Henry Campion, a beer brewer used it and Abraham, his son, now possesseth it.'

Taken over by Calvert & Co. in 1730, it was brewing 53,000 barrels per year by 1748. In 1810, the City of London Brewery absorbed Calvert's long-established Peacock Brewery in Whitecross Street, off Old Street, which then closed. The company moved to the Swan Brewery in Walham Green, Fulham, in 1914, at

Cannon Brewery, St John Street, Clerkenwell.

which time the Hour Glass Brewery was converted to warehouses. All brewing ceased at Walham Green in 1936.

Arthur Guinness began brewing in Dublin in 1859 at the St James's Gate Brewery. The company was enormously successful – by the 1930s they were the seventh largest company in the world. What became the largest brewery in the UK was opened in Park Royal in 1933. Guinness is now brewed in fifty countries. However, Park Royal closed in 2005, and all Guinness sold in the UK and Ireland is now brewed in Dublin.

Distilling

Gin was first distilled in the seventeenth century by the Dutch chemist Professor Franciscus Sylvius. He prepared it from rye and then rectified (redistilled) the spirit and added juniper as a flavouring. It became known as genièvre (juniper in French), from which the shortened name gin is derived. It had a reputation as a cure for many ills, including lumbago, gallstones, gout, and kidney ailments. Gin came to the attention of British soldiers fighting in Holland; they drank it before battle – hence the term 'Dutch courage'. But it was the Glorious Revolution of 1688 and the coming to Britain from Holland of the joint monarchs William and Mary that assured its popularity in England. Almost at once it became the drink of the poor.

Gin shops sprang up everywhere, especially in London. Consumption rose at an alarming rate. In 1690, 500,000 gallons were distilled but by 1729 this had increased to 5 million gallons. And it was often of poor quality, sold in dram shops on every street corner – over 7,000 in London alone. Daniel Defoe wrote how the distillers had 'found out a way to hit the palate of the poor by their new fashioned compound waters called geneva; so that the common people seem not to value the French brandy as usual, and even not to desire it'. Hogarth's painting 'Gin Lane' vividly illustrates the social problems caused by gin, as did Charles Dickens who wrote of the 'filth and squalid misery' outside the gin palaces of Covent Garden and St Giles. A notice outside a gin shop in Southwark read: 'Drunk for 1d, dead drunk for 2d, clean straw for nothing' (straw was provided in the cellar for those dead drunk to sleep it off). People began to speak out against the evils of gin and it was obvious something had to be done. Alexander Pope wrote:

Vice thus abused, demands a nation's care
This calls the Church to deprecate our sin
And hurls the thunder of the laws on gin.

A series of Acts of Parliament, beginning in 1729, were placed on the statute book. In 1729, a tax was imposed – only to encourage the production of poor-quality spirit. Then, in 1733, it became unlawful to sell spirits in the street, but all this did was increase the number of gin shops. The Gin Act of 1736 placed high taxes on retailers, prompting the London Mob to riot. Prohibition was obviously not working and eventually a policy was adopted to price spirits at a sufficient level to discourage immoderate drinking. Duties were imposed on distillers and retailers licensed under the supervision of a magistrate. As we saw earlier, it was only when the duty on beer was withdrawn, in 1830, that gin gave way to beer as the drink of choice of the poor.

Philip Booth & Co.

In his book *The Kindred Spirit*, Lord Kinross describes how gin was produced at the firm of Philip Booth & Co. of 55 Turnmill Street, Clerkenwell. Raw materials are maize and barley. Maize is ground into grist in milling machines, while barley is soaked in water and converted to malt. Both are then mashed together in mash tuns where starch is converted to sugar. The resulting liquor (wort) is drained into other tuns and yeast added to convert the sugar to alcohol. The product is distilled in wooden stills and the alcohol evaporated off and condensed. The liquid spirit is then redistilled twice, flavours added and then stored in oak vats before bottling.

The first record of the firm of Philip Booth & Co. is in a 1778 directory of merchants, but it is certain that Booth's were well established before then. Philip

Booth was plainly a man of means, living in fashionable Russell Square before moving to Crouch End and finally Waltham Abbey. The company expanded their operations to Stanstead Abbotts under the direction of Philip's son, William. But it was another son, Felix, who moved the firm forward in London. He bought up Hazard & Co.'s Red Lion Brewery in Brentford and built a distillery.

Up the river from Brentford is Bushy Park and there lived the Duke of Clarence – soon to be William IV. Felix made his acquaintance and the duke paid a visit to the distillery, noting he was 'pleased to express his approbation of the whole arrangements' and 'was particularly struck with the novel appearance of the feeding arrangement for so many oxen'. (Booth's kept cattle – feeding them on residues of the distillation process – a profitable sideline.) As well as distilling gin, Booth's brewed beer as well, and William IV granted the firm the privilege of calling the brewery the Royal Brewery.

But with the arrival of the age of gas, Booth was quick to diversify into this new industry, becoming chairman of the Brentford Gas Works, sited next door to his distillery. He also had a philanthropic interest in science. In 1829, he sponsored Captain James Ross' paddle steamer, *Victory*, in its search for the North West Passage. The voyage resulted in the discovery of the true position of magnetic north and the naming of the peninsula Boothia after Felix Booth.

Booth's was now the biggest distiller in England. It bought the cognac brandy distillery in Albany Street, Regent's Park, and then, in 1923, the company acquired the firm of John Watney & Co. in Wandsworth. Booth's offices are housed in an Edwardian building designed by E.W. Mountford. The lower part is faced with granite from Dartmoor, while the upper part is of red brick and Portland stone. There are reliefs by F.W. Pomeroy. In the 1950s, a new distillery was built on the corner of Clerkenwell Road and Red Lion Street.

Relief on the premises of Booth's Gin showing harvesting for gin production, Britton Street (formerly at Turnmill Street).

Gilbey

In 1857, Walter and Alfred Gilbey rented some murky cellars on the corner of Berwick Street and Oxford Street. Together with Henry Gold they began to import and sell South African wines. Trade boomed and in 1860 they acquired larger premises in Titchfield Street and soon had offices in The Pantheon in Oxford Street. In 1869, Gilbey's leased a railway shed in Camden Town from the London & Birmingham Railway Co. They added the Camden flour mill, recently destroyed by fire, and in 1872 built a distillery on the site. Later, the company took over the well-known Roundhouse. In 1936, Marks & Spencer bought The Pantheon site (their store is there to this day) and Gilbey's took the opportunity to build new offices in Camden.

Beefeater

James Burrough was born in Devon. He qualified as a pharmacist and took his skill to America in 1835. Back in England, in 1863, he purchased a distillery in Cale Street, Chelsea, for £400. The distillery dated from about 1820 and was owned by the Taylor family. An early customer was the famous Piccadilly store of Fortnum & Mason. Then, between 1876 and 1879, Beefeater Gin was launched, taking its name from the Yeoman Warders (Beefeaters) at the Tower of London. Expansion prompted a move to Hutton Road in Lambeth in 1908, and then to Kennington on a site in Montford Place previously occupied by Hayward's Military Pickle Factory.

Gordon's

The story of Gordon's Gin begins with Alexander Gordon, who opened a distillery in Southwark in 1769. Seventeen years later, he moved to a distillery in Goswell Road, Clerkenwell. In 1898, Gordon's amalgamated with Charles Tanqueray & Co. Over the years Gordon's were awarded many royal warrants, including those by George V, George VI, the late Queen Mother and the present Queen. On 11 May 1941, the Goswell Road distillery was bombed. Rebuilding was complete in 1957 and 'Old Tom', one of the original stills, was reinstated. In 1984, all production moved to Basildon in Essex.

J. & W. Nicholson & Co.

Nicholson's was founded in the 1730s. John and William Nicholson joined their cousin John Bowman and set up as distillers, and wine and brandy merchants. By 1809, the Nicholson brothers were operating alone in Woodbridge Street, Clerkenwell. In 1828, they moved to a new distillery in St John Street. They also had premises at Three Mills, Bromley-by-Bow. The tidal mills at Three Mills lie on the River Lea and are still standing for us to enjoy today. There were many other mills powered by the Lea, including Temple Mill, Abbey Mill and Pudding Mill, and all belonged in the Middle Ages to Stratford Langthorne Abbey. The House Mill, built in 1776 by Daniel Bisson, was the largest tidal mill in the world. Nicholson & Co. took possession in 1872 and the mill was used for grinding grain for gin production until 1940. Adjacent is the Clock Mill, rebuilt by Philip

Metcalfe in 1815. Distilling ceased in 1970 when the company sold up to the Distillers Co. Three Mills was then used as a bottling plant for Bass Charrington.

25 The Building Industry

Nicholas Barbon

Nicholas Barbon was the son of 'Praise-God Barbon', leather seller at the sign of the 'Lock & Key' in Crane Court, Fleet Street, and eccentric MP in the interregnum. Barbon, the son, studied at Leyden in the Netherlands, obtained a diploma as a Doctor of Medicine and is best known to us – or at least those concerned with the development of London – as a speculative builder. Barbon became MP for Bramber in Sussex in 1690, but his life can best be summed up by the comment of an acquaintance: 'All his aim was profit.' The impetus for Barbon was the aftermath of the Great Fire: London was rebuilt, it expanded westwards, and Barbon stepped in and seized his chance.

Of the many grand houses lining the Strand, the first to catch Barbon's eye was Essex (or Exeter) House, previously home to Elizabeth I's ill-fated favourite, Robert Devereux, Earl of Essex, and then in the hands of the Duchess of Somerset. In 1674, Barbon bought Essex House, knocked it down and built 'houses and tenements for taverns, alehouses, cookshoppes, and turned the gardens adjoining the river into wharves for brewers and woodmongers [coal merchants]'. Essex Street, Little Essex Street and Devereux Court are still there today.

Next was York House, a few hundred yards to the west. It was formerly the London home of the Archbishop of York and later George Villiers, Duke of Buckingham. Barbon pulled it down and rebuilt, but the grand old duke insisted that every part of his name be remembered in Barbon's new streets – hence present-day George Street, Buckingham Street, Duke Street, Villiers Street and even Of Alley. In this area, south of the Strand, Barbon built well. One of his streets in the York Buildings development was termed 'a handsome street with good houses well inhabited'.

Holborn and Gray's Inn were ripe for development, and Barbon soon turned his attention in that direction. The land included Gray's Inn Fields, Red Lion Fields and Lamb's Conduit Fields, together with the Red Lion Inn, famous as a holding place for the bodies of Cromwell, Ireton and Bradshaw on their journey to Tyburn. (A successor inn is on the site today.) In Red Lion Fields, Barbon met opposition from the lawyers of Gray's Inn. One, Narcissus Luttrell, related in his diary that, on 10 June 1684:

> Dr Nicholas Barebone, the great builder, having some time since bought the
> Red Lyon Fields near Gray's Inn Walks to build on, and having for that purpose

employed several workmen to goe on with the same, the gentlemen of Gray's Inn took notice of it, and thinking it an injury to them, went with a considerable body of 100 persons; upon which the workmen assaulted the gentlemen and threw bricks at them; so sharp an engagement ensued, but the gentlemen routed them at the last and brought away one or two of the workmen to Graie's Inn; in this skirmish one or two of the gentlemen and servants of the House weer hurt, and several of the workmen.

The riot came to the attention of the Attorney General who directed 'the suppression of Dr Barbon and his men from committing any insolence in their late riotous meeting in Red Lion Fields and to prevent them from annoying His Majesty's subjects'.

Matters did not rest there. The lawyers complained that Barbon's houses would be 'a nuisance to divers persons of honour and quality'. This would include the King, whose private highway to Theobald's and thence to Newmarket would be impeded. Barbon was not deterred – he simply threatened to break down the gates and proceeded with his development regardless.

As well as the Red Lion Square development, Barbon built houses in Bedford Street, Princes Street, Theobald's Row, Eagle Street and Lamb's Conduit Street. In the latter he disturbed the water supply from the conduit to Christ's Hospital School in the City. The school governors insisted that:

> … it was used to make broth for 400 children each day and even in the depths of winter never failed. [Also] the said water by only outward washing of the body curreth the scorbutic humours and some have aver'd the leprosy, therefore by no means may it be lost or parted with.

Like many a sharp businessman, Barbon tried every ruse in the book to avoid paying his debts. Sometimes, rather than paying a bricklayer's bill he would offer him a house instead, and if called a rogue and a cheat he would merely smile. Throughout his life Barbon came in for much criticism from his contemporaries. He was quick to reply with a pamphlet entitled *An Apology for the Builder*, pointing out:

> The natural increase of mankind is the cause of the increase of the city, and that there are no more houses built each year than are necessary for the growth of the inhabitants, as will appear by the number of apprentices made free, and marriages every year in the city … there ought to be 1000 houses at least built every year for the 9000 apprentices that come out of their time, and the 10,000 weddings to have room to breed in.

Barbon's philosophy anticipated that of Adam Smith. Like Smith, Barbon was a firm believer in free trade: 'The prohibition of any foreign commodity doth hinder the making of much of the nature.' In 1690, he wrote another pamphlet, *The Discourse of Trade*:

The expense that chiefly promotes trade is building, which is natural to mankind, being the making of a nest or a place for his birth, it is the most proper distinction of riches and greatness, because the expenses are too great for mean persons to follow ... Building is the chiefest promoter of trade, it imploys [*sic*] a greater number of trades and people than feeding or clothing.

By 1694, four years before his death, Barbon got himself into serious financial difficulty. It was left to a business associate to sum up his life and character:

By contrivance, shifting and many losses, he kept his wheel turning, all the while lived splendidly, was a mystery in his time, uncertain whether worth anything or not, secured at last a Parliamentman's place, had protection and ease, and had not his cash failed, which made his work often stand still and so go to ruin, and many other disadvantages grow, in all probability he might have been as rich as any in the nation.

Thomas Cubitt

Thomas Cubitt can, with all justification, claim to be London's most prolific builder. Born in 1788, just outside Norwich, he trained as a carpenter and served as ship's carpenter on a voyage to India. He set up in business in Holborn and was soon joined by his brother, William. In 1815, the pair moved to Gray's Inn Road. In contrast to other builders, rather than taking on workmen – bricklayers, plasterers, carpenters and masons – in an ad hoc fashion, they employed a full-time workforce and kept them busy. To achieve this they needed a constant supply of work and that is exactly what they got. Starting in a modest way, the Cubitts had small developments in Highbury, Stoke Newington and Barnsbury, and built the London Institute in Finsbury Circus. Next came scores of houses on the Duke of Bedford's estate in Bloomsbury, including the west side of Tavistock Square, Woburn Place and Gordon Square. Thomas Cubitt is best known, however, for his vast development in Belgravia and Pimlico, employing George Basevi as architect. In 1827, the partnership between Thomas and William broke up; William remained in Gray's Inn Road, while Thomas moved to a building works in Pimlico at Thames Bank.

A firm favourite with the royal family – they referred to him as 'Our Cubitt' – Thomas was awarded the contract to carry out Edward Blore's and James Pennethorne's extension to Buckingham Palace. Taking great pride as a builder rather than an architect, he was also responsible for the vast mansions at Albert Gate. They were sometimes known as Malta and Gibraltar because 'they would never be taken'. Everything that Thomas Cubitt did was in 'a magnificent style': land was drained properly, sewers were installed, roads were made and his estates were well lit. Such ancillary work was an essential part of the developer's job. William Cubitt, meanwhile, built Fishmongers' Hall, Covent Garden Market,

Statue of Thomas Cubitt, Denbigh Street, Pimlico, by William Fawke.

Euston Station and Shadwell Basin in London Dock. He also acquired land on the Isle of Dogs from the Glengall Estate, and as well as his own works there built factories and houses for the working classes. The area became known as Cubitt Town. There was also a younger brother, Lewis, who designed King's Cross Station.

Other Builders

In Notting Hill, Charles Henry Blake, recently retired from the Indian Civil Service, was a developer, landowner, financier and builder. In five years he spent £116,000 putting up the Kensington Park Estate. Unfortunately for Blake, his builders turned out to be unreliable, and it was only when the railway came in 1864 that his schemes were saved from failure.

By the mid-nineteenth century an enormous building programme was under way in London. There was a mass exodus from the City – in 1861, there were 113,000 people living there; ten years later, the population had fallen to 76,000. Those moving out needed to be rehoused somewhere in the burgeoning suburbs. Probably as many as 10 per cent of the male population were employed, either directly or indirectly, in the building industry. There were hundreds of building firms tendering for hundreds of contracts, but of these, Summerson writes in *The London Building World of the Eighteen-Sixties* that in 1861, only thirty-nine tendered for jobs of over £10,000 and only seven of these for more than one large contract at a time. The seven – presumably the largest firms in London – were Mansfield, Myers, Willson, Lawrence, Kirk & Parry, Hill Keddle and Lucas.

Little is known of these men who built Victorian London. George Myers tendered for forty-four contracts in 1865 and got five of them. He was born in Hull at the beginning of the nineteenth century and worked as a mason at Beverley Minster. It was here that he met Augustus Welby Pugin. Both men were engaged in examining out-of-the-way parts of the Minster. Pugin was master of the Gothic style and Myers became a willing convert. He set up at Ordnance Wharf in Belvedere Road, Lambeth, in 1844 and did much work for Pugin.

Charles Lucas was born in Norwich and started in business there with his brother, Thomas. He joined the firm of Samuel Morton Peto – well known as one of the builders of the Royal Victoria Dock. Peto's business failed in the mid-1860s and Lucas took it over, moving to Lambeth, next door to Myers. He was later to build for Sir Charles Barry and made a fortune from his work at Covent Garden Theatre, the International Exhibition of 1862, the Royal Albert Dock and the new cut-and-cover Underground railway. Nearby was Lawrence at Pitfield Wharf, Lambeth (now the site of the National Theatre).

Dove Brothers were founded by William Spencer Dove, who was born in Sunbury in 1793. He began in the 1820s, doing small building jobs in Islington, but soon began building houses. The firm's first yard was between Moon Street and Studd Street, Islington, and in 1901 they moved to nearby Cloudesley Place. Dove Brothers had an excellent reputation and between 1858 and 1900 built 130 churches, many public buildings, shops, offices and houses.

Allied to the building industry were timber merchants. Charles Frederick Anderson began as a timber dealer in 1863. He operated from Essex Road and, as well as supplying timber to the building trade, had contracts with Elstree Film Studios where scores of film sets were constructed each year.

26 The Coal Trade

The Industrial Revolution was powered by coal. It fuelled London's early steam engines, then the boilers of its power stations, and was the raw material in all the capital's many gasworks. There is no mention of coal in Domesday Book, William the Conqueror's great survey of 1086; at that time wood was preferred. But when a royal decree discouraged burning wood – it depleted forests of trees intended for shipbuilding – Londoners turned to coal to warm their homes. It is hardly surprising, therefore, that the great majority of the ships entering the Thames in the eighteenth, nineteenth and twentieth centuries were from the north-east and carried coal.

By the thirteenth century, coal mined in Northumberland was being imported on a regular basis to London. It was known as seacoal and its main purpose was for lime burning. Not that coal was immediately popular. In 1273, it was forbidden to be burnt in the City because of the noxious fumes emitted. Edward I soon rescinded the decree, only for Londoners to petition against it again in 1306. And it was not long before it was taxed. In the reign of Edward I, a tax of 6*d* was imposed on every seacoal vessel passing under London Bridge. Most came from the 'good men of Newcastle', who shipped coal from the Tyne. In the mid-sixteenth century a mere handful of ships arrived; by 1705, the number had risen to 600, and in 1805 as many as 4,800 colliers were recorded entering the port.

Trade was disrupted for a while in the Civil War; Newcastle upon Tyne was loyal to the King but London was on the side of Parliament. As a result, in 1642, Parliament banned all trade with Newcastle, the unfortunate outcome of which was a severe shortage of coal in London, alleviated only somewhat by alternative supplies from Sunderland and Blyth.

After normal supplies were restored, it was the Dutch Wars that then interfered with the passage of ships from Newcastle. Samuel Pepys noted in his diary in December 1665:

> In much fear of ill news of our colliers. A fleet of two hundred sail and fourteen Dutch men-o-war between them and us: and they coming home with small convoy: and the City in great want, coals being at three guineas per chaldron.

(A chaldron is equivalent to about 53 hundredweight.)

When the Great Plague was raging, smoke from coal was believed to prevent the dread disease. Daniel Defoe records in *A Journal of the Plague Year*:

> Great quantities of coals were then burned, even all the summer long and when the weather was hottest, by the advice of the physicians, for the heat of the fires tended not to nourish but to consume all those noxious fumes which the heat of the summer generated, the bituminous substance and sulphurous and nitrous particles assisting to clear and purge the air. The latter opinion prevailed at that time and, as I must confess, with good reason, and the experience of the citizens confirmed it, many houses that had constant fires kept in the rooms having never been infected at all. By order of the Lord Mayor, large public fires were made in certain streets, which must have cost the city about 200 chalder of coals a week, but as it was thought necessary, nothing was spared.

In the medieval era, coal imports were regulated by the Woodmongers' and Coal Sellers' Livery Company, first recorded in 1376. At a similar time, a set of officials called Coal-Meters were established. It was their job to weigh or measure the coal, a task none too easy because of its bulk. They also helped to gather taxes.

The cost of coal – inflated by tax – and the fact that many could not afford to buy it, was a constant complaint. In 1792, a pamphleteer wrote: 'Damp bedrooms, damp beds, damp cloths, … many are devoured by rheumatisms, colds, fevers and consumptions, … every bed room ought to have once a day a good fire.'

The money raised from coal tax was put to many purposes. It paid for one of the campaigns of the Black Prince in the Hundred Years War. St Paul's Cathedral

THE OLD COAL EXCHANGE

The Coal Exchange, Lower Thames Street, 1830.

and many other City churches were built on its proceeds, as well as the first
Waterloo Bridge, Holborn Viaduct and the Thames Embankment. Import duties
on seacoal finally ended in 1831. After the Great Fire, the City of London taxed
coal imported by land to London and sold within a 25-mile radius of the Royal
Exchange. Coal tax posts, marking out the boundary, are a familiar sight to this
day. The duty finally elapsed in 1889.

One year after the Great Fire, the Woodmongers lost their charter because
of alleged profiteering, their master, Sir Edmund Berry Godfrey, getting himself
murdered soon afterwards. Control then passed to the Watermen and Lightermen,
who amalgamated in 1760. About 1,400 colliers were bringing their seacoal into
the port at the beginning of the eighteenth century. Each had a crew of ten or
twelve men and a capacity of 140 tons. Every year 335,000 chaldrons of coal were
imported, increasing to just over a million chaldrons by 1800.

Coal wharves were situated along the banks of the Thames, with many of them
in Thames Street, Bankside and Shadwell. The well-known 'Prospect of Whitby'

THE PROCESSION SEEN WESTWARD FROM WATERLOO-BRIDGE.

Procession on the River Thames to mark the opening of the new Coal Exchange, 1849. Note the incomplete tower of Big Ben.

public house commemorates a collier of that name that used to discharge Whitby coal at Shadwell. There was money to be made, and many entered the trade. Buyers purchased from the ships' masters via middlemen, so-called coal factors, and deals were done in the taverns of Thames Street. 'The Dog Tavern' stood between St Mary at Hill and St Dunstan's Hill, and was described by Stow as 'a very handsome and genteel place and well inhabited'. The 'Newcastle Coffee House' and 'The Sun Tavern' were other places where business was conducted. In time, the trade became more organised and the Coal Exchange was founded, built in 1768 in Thames Street. A new Exchange was opened in 1849 by the Prince Consort with all the usual ceremony. A splendid and colourful procession on the Thames preceded the event, held in the presence of the Princess Royal and the Prince of Wales.

By the early twentieth century, the discharge of coal by the fleets of colliers had become highly structured. W.J. Passingham in *London's Markets* tells us that in the 1930s each collier had its own identification mark, so that when a vessel entered the river it received specific instructions about where to discharge its cargo. Information from *Lloyd's List* explains:

When a London collier starts upon her life's work she is assigned a definite and individual number by the Coal Factors' Society. She is identified by this number by the representative of the Society who keeps watch in the Collier Signal Station at Tilbury Fort. [When the ship has passed by] a telephone message is relayed to all concerned – at Nine Elms or Fulham or wherever it might be – that number xxx is on her way.

With the coming of the railways, an alternative means became available to import coal. To begin with, the railway companies were reluctant to carry coal – they reasoned it would put off the travelling public. But by the mid-nineteenth century, railway transport began to compete with traditional seaborne vessels. Furthermore, rail supplies were taken from mines in the East Midlands, Leicestershire and Warwickshire, posing a threat to supplies from Newcastle. The wealthy Newcastle coal owner Sir Charles Palmer responded with the remark, 'Make steam compete with steam,' and it was Palmer who built the first iron screw steam collier, the *John Bowes*, at a cost of £10,000, capable of carrying a cargo of 650 tons of coal. The old sailing colliers took a month to reach London but Palmer's new steamship made the journey in just five days. His vessels discharged at Poplar Dock, where he had installed a hydraulic power plant to offload the coal to waiting railway wagons and so distribute it all over London.

27 Food

Food was produced locally in medieval England. A tenant farmer supplied food for his own requirements; any left over he would sell to his neighbours. This kind of home producer arrangement was perfectly adequate for a rural society, but broke down as the population increased and more people began to live in towns, the Industrial Revolution finally sounding its death knell.

By the eighteenth century, London was surrounded by market gardens. And it is to Huguenot refugees that we owe a debt of thanks for their introduction. Samuel Hartlib, in his *Legacy of Husbandry*, wrote in 1651: 'About fifty years ago, this art of gardening began to creep into England, into Sandwich and Surrey, Fulham and other places.' The Huguenots planted cabbages and cauliflowers and sowed turnips, carrots, parsnips, peas and rape, 'all of which at that time were great rarities, we having few or none in England, but what came from Holland and Flanders'; and 'the rich garden grounds, first planted by the Flemings, still continue to be the most productive in the neighbourhood of the Metropolis'. London's soil – predominantly gravel – was quite unsuitable for growing vegetables, so the Huguenots dug it up and, as the Venetian Ambassador to London noted, 'filled the holes with the filth of the city, an excellent manure as rich and black as thick ink'. There were market gardens in many of London's inner suburbs: Battersea, famous for its asparagus; Clerkenwell; Shoreditch; Stepney; Tothill Fields in Westminster; Fulham; and Cherry Garden, Bermondsey. And there was money to be made, for Thomas Fuller, in *Worthies of England* in 1660, tells of the 'incredible profit of digging the ground'.

Smithfield Market

Cattle were imported to Smithfield – the oldest and largest wholesale meat market in London. It was described by William Fitzstephen, in 1174, as a 'smooth-field [hence Smithfield] where every Friday there is a celebrated rendezvous of fine horses to be sold and in another quarter are placed vendibles of the peasant, swine with their deep flanks, and cows and oxen of immense bulk'. Edward III granted the market a charter in 1327, and soon drovers were bringing tens of thousands of cattle, sheep and pigs from far and wide, entering the market from Cowcross Street or St John Street. The City Corporation collected dues for all animals, which would be slaughtered there and then. Not surprisingly, conditions were appalling. Charles Dickens describes the scene in *Oliver Twist*:

> The ground was covered, nearly ankle deep, with filth and mire; a thick steam perpetually rising from the reeking bodies of the cattle. The unwashed, unshaven, squalid and dirty figures constantly running to and fro, and bursting in and out of the throng, rendering it a stunning and bewildering scene, which quite confounded the senses.

It was not uncommon for cattle to run free and run riot in local shops – hence the saying 'like a bull in a china shop'. The City Corporation had to do some-thing about it and so, in 1852, the Smithfield Market Removal Act was passed and the live cattle market moved to Caledonian Cattle Market at Copenhagen Fields, Smithfield reverting to a wholesale market.

The market was given an enormous boost with the coming of the railway, allowing meat to be imported directly. Shortly afterwards, in 1865, Sir Horace Jones built the market building we see today. It was on two levels. Beneath was a basement where meat was unloaded off railway wagons and above was (and is) the market hall itself. The building – influenced in its design by Joseph Paxton's Crystal Palace and opened in 1868 – is 630ft in length and is divided by the cen-tral arcade (the Grand Avenue). In 1875, a poultry market – the London Central Poultry and Provision Market – was built. It burnt down in 1958 but was rebuilt. The Beeching Axe brought rail deliveries to an end and the enormous under-ground sidings have now been converted into a car park.

The market is administered by the Corporation of London. Vast juggernaut lorries, many from continental Europe, arrive throughout the day and unloading begins at midnight. There are definite lines of demarcation. 'Pullers-back' haul the carcasses to the tailboard of the lorry, then 'pitchers' load the meat onto their shoulders and run (the well-known Smithfield Shuffle) to offload it to hooks at the tenant's stall. Tenants rent the market stalls, where 'humpers' or 'shopmen' weigh the meat and put it on display. Trading starts at five in the morning and all sales are negotiated freely between seller and buyer. Once a deal is done, 'bummarees' carry the meat to the buyer's waiting van. By 8 a.m. all is done and all that remains is for the City Corporation cleaners to clear up the mess. Large supermarkets have taken much trade away from Smithfield and now the market sells mainly to hotels, restaurants and local butchers. In recent years, the market has had a £70 million refurbishment to enable it to comply with current hygiene standards.

Billingsgate Market

Early morning is the best time to view the action at Billingsgate Market and this means between 4.30 a.m. and 8.30 a.m. The converted transit shed on the Isle of Dogs is the successor to the famous Billingsgate Fish Market in Thames Street in the City. The name probably derives from Beling the Saxon, who was a local landowner. In 1327, Edward III granted a charter that prevented rival markets setting up within 6.6 miles of Billingsgate. (This was the distance that, within one day, a man could reasonably walk to Billingsgate, sell his goods, and then return home.) In those early days, Billingsgate traded in all sorts of commodities, including coal, corn, iron, pottery, salt, wine and fish, and it was not until about 1699 that the market was restricted solely to the fish trade.

It was particularly notorious for the bad language of its porters, who all wore special hats based on those worn by the longbow men at the Battle of Agincourt. Sir Horace Jones' fine building was converted by Richard Rogers for the financial services industry in 1982 and the fish market moved to its present site on the Isle of Dogs. The original Billingsgate Bell survives at the new market and there is a copy of the clock from the old market hall. The market is administered by the Corporation of London and early every weekday morning lorryloads of fish arrive from all points of the globe. The new building has a large floor space and the method of trading follows a well-established pattern. Samples of fish are brought to the market hall by licensed porters and put on display. This is known as 'shoring in'. Buyers then place their orders and the same porters collect the sales orders from the suppliers and 'barrow' them to the buyers' vehicles. For this service, the buyer pays the porter 'bobbin money'. Every year, as rental payment, a gift of fish is given to the Mayor of Tower Hamlets who then distributes it to the residents of old people's homes.

Covent Garden

It was Charles II who granted William Russell, 5th Earl of Bedford a charter to open a market for horticultural produce in 1670. The market grew and, in 1830, Charles Fowler built the range of buildings – now given over to small shops, restaurants and open-air entertainment – we see today. The place obviously impressed Charles Dickens:

> Covent Garden on market morning is wonderful company … at sunrise in the spring or summer, when the fragrance of sweet flowers is in the air, overpowering even the unwelcome streams of last night's debauchery, and driving the dusky thrush, whose cage has hung outside a garret window all night long, half mad with joy.

In 1918, the 11th Duke of Bedford sold to a property company and later, in 1961, the Covent Garden Market Authority came into being. But by then the future of Covent Garden was in doubt. The congestion caused by large lorries offloading in the narrow streets of central London had become intolerable. A move was therefore made in 1994 to a 68-acre site at Nine Elms, where as many as 2,000 lorries could be accommodated. Now known as New Covent Garden Market (NCGM), it is the largest fresh produce market in the UK, with over 200 businesses occupying the site and employing 2,800 people. The railway tracks from Waterloo cut through the site – on one side is the fruit and vegetable market, and on the other the flower market. Goods are offloaded from lorries to fork-lift trucks and conveyed to tenants' stalls. Most tenants act for the grower and sell at the best price they can get, taking a commission for themselves. However, some direct selling does take place from lorries. There is activity from about 11 p.m. and buying starts at 4 a.m., thereby ensuring that buyers – greengrocers, restaurants and hotels – have fresh produce for the day ahead.

Spitalfields Market

The name comes from the medieval hospital and priory of St Mary Spital. In 1682, John Balch was granted a charter by Charles II to hold a market on Thursdays and Saturdays. Then, in 1876, Robert Horner bought the site, completed new buildings and took advantage of the 1682 charter to stop rival fruit and vegetable markets from trading without paying dues. After a bitter legal battle, the City Corporation took control of the market in 1902 and expanded it. It was still going strong in the 1980s with 150 wholesalers. In contrast to other markets, there were no independent freelance porters at Spitalfields; each tenant employed his own porters, clerks and salesmen. In 1991, the market moved to a 31-acre site in Leyton.

Borough Market

The first Borough Market was held on the London Bridge built in 1014 by the Danish King Cnut. But the obvious congestion forced a move to the nearby

Borough Market, Southwark.

grounds of the Hospital of St Thomas, south of the river in Southwark. Later, Edward III, in a series of charters, granted land in this part of Southwark to the City Corporation. The Bridge House Estates (the body responsible for the upkeep of London Bridge) financed a separate market which extended down Long Southwark (present-day Borough High Street). So now there were two markets in the area, forcing the one operating from the hospital to close. In 1755, the City Corporation, worried about congestion at the entrance to London Bridge, tried to force the second market to close.

Local people would have none of it and under the guidance of the church-wardens of St Saviour's Church (Southwark Cathedral) purchased land – previously the churchyard of the demolished church of St Margaret, and known as The Triangle – and moved the market there, where it remains to this day. The market survived the disruption caused when the railway was extended from London Bridge to Charing Cross and Cannon Street. The buildings were designed by H. Rose in 1851 and E. Habershon in 1863.

'London's Breakfast Table'

Food has always been imported to London, at first supplying local retail markets. The Romans had a market or forum between Gracechurch Street and Leadenhall Street. Westcheap (Cheapside) and Eastcheap were the main retail markets in medieval London ('cheap' means market place in Old English) and a glance at the names of surrounding streets – Bread Street, Milk Street, Poultry, Honey Lane – tells us immediately what was sold there. Salted butter was transported by boat from Northumberland and Carmarthen, and large quantities of bacon, eggs, lard and ham were imported from Ireland.

But London's population was increasing rapidly and, because home supply was insufficient to meet demand, increasing amounts of food began to be imported from abroad. Customs dues were levied on all imports which, since the Tudor era, were by law landed at the Legal Quays on the north bank of the Thames between London Bridge and the Tower. As time went by, more and more vessels entered the Pool of London to offload at the Legal Quays, and not just food, for all manner of goods had to be discharged there. Congestion became intolerable and, in 1789, so-called Sufferance wharves were established elsewhere where goods of lesser value, such as foodstuffs, could be offloaded 'under sufferance'. Many Sufferance wharves were on the south bank of the Thames in Bermondsey. Until the 1960s, the entire stretch of river from London Bridge to Tower Bridge was dominated with wharves belonging to the Hay's Wharf group. Overhanging cranes unloaded New Zealand dairy products from clusters of lighters, and in the nineteenth century and before, banks of sailing ships waited to unload tea from China. The area became known, with every justification, as 'London's larder', or if not that, 'London's breakfast table'. In the twentieth century, 75 per cent of the capital's butter, cheese and canned meat were stored here!

Hay's Wharf was one of the oldest wharves on the river. It all started in 1651, when Alexander Hay took a lease on land previously occupied by a brewhouse. Hay and his sons, John and Joseph, were carpenters and they made pipes for Hugh Myddelton and his New River Company to convey drinking water. It was not long, however, before imported goods began to be landed at what was then known as Pipe Borer's Wharf. There were other wharves as well, such as Topping's Wharf and Pickleherring, which took its name from the pickled herrings landed there from Norfolk and bound for the tables of the monks of Bermondsey Abbey. In 1793, Topping's Wharf came up for sale by auction. The sale notice described the premises in glowing terms:

> The warehouses are very substantially built, in good repair, and are capable of housing 1500 hogshead of sugar. There is space for two tiers of vessels to lie abreast next to the wharf, with a considerable and profitable trade belonging to the premises which from their situation and accommodation they will always command.

In 1838, upon the death of Francis Theodore Hay, ownership of Hay's Wharf passed to John Humphrey. He was soon to ask William Cubitt to build a new wharf and at the same time construct a wet dock, Hay's Dock. A vast array of goods were imported into Hay's Dock – salt, sugar, raw coffee, tea, potatoes, dried fruit, rice, hemp, etc. Opened in 1857, it has now been filled in and the site is occupied by Hay's Galleria. In 1867, the Hay's Wharf group pioneered the use of cold storage to preserve perishable goods and, in 1879, the first consignment of frozen meat was landed.

Butler's Wharf is downstream of Tower Bridge and in its day was London's most concentrated series of warehouses. Its early history is shadowy. John Roque's map of 1746 shows timber wharves in the area, and later there were ropeyards, granaries and shipyards. In 1794, a Mr Butler in partnership with a Mr Holland operated as wharfingers and, in 1872, Butler's Wharf was established as a registered public company. The manager was Henry Lafone, brother of Albert, a local politician who gave his name to nearby Lafone Street. Butler's Wharf was famous for handling tea and by 1950, 6,000 chests a day were being imported. But tea was by no means its only commodity – there was rubber, canned meat and fish, rice, sugar, honey, pepper and much more.

Life for the docker was tough. Louise Roche has written a social and economic history of Butler's Wharf and interviewed many dockers who gave her their own experiences of working there. Life was hard and the same system of 'call on' (casual work) was employed as elsewhere in the docks. One warehouse man recalls that 'if there was a boat in, a bloke called out so many and they didn't want no more'. Also, in the early twentieth century, 'they used to work all day for half a crown and sometimes they never got a day's work they only got half a day'. It was sometimes worse than that: 'They'd call them off and pay them off when they wanted to. They done two hours work, they got paid two hours money and that was it.'

Butler's Wharf closed in 1972 and apart from its occasional use as a setting for film crews, including *Doctor Who*, it lay idle until restored as flats by Conran Roche.

There were many stages between the farmer in his field and the hungry Londoner at his dining table – shippers, agents, wholesalers and retailers. To regulate trade, in 1866, the Home & Foreign Produce Exchange was opened in Hibernia Chambers, Southwark. Its aims were 'to regulate the course of trading between producers and others on just and equitable principles to provide useful and suitable forms of contract to collect and disseminate information'. Importers or agents would take orders from wholesalers. Food would be held either in cold storage on board ship or in one of the warehouses in Tooley Street. A market price would be fixed by supply and demand. No foodstuffs entered the exchange, so if buyers wished to sample the goods they would be asked to do so at the warehouse. Buyers of butter would insert a long, elongated 'iron' into a cask of butter, withdraw it and bite a piece off. This rather unsavoury practice came to the attention of the local medical officer, prompting him to query samplers '[putting] the butter back in the cask again with the marks of their teeth on it and anything that may come from their lips and teeth'.

Many firms operated in Tooley Street. Mills & Sparrow had built their headquarters, Colonial House, by 1904 and were joined by Henry Lane & Co. at 59 Tooley Street. Hugh Barty-King in *Making Provision: A Centenary History of the Provision Trade* tells how George Sims, in his *Living London*, describes the scene in 1903:

> … and there are carts and railway vans bringing in or taking away hams, bacon, eggs, butter and cheeses of foreign origin; glance inside Hay's Wharf, at the two towering rows of warehouses whose restless cranes are whirling goods up, from or down to the tangle of vehicles below that are unloading or loading; go round behind the warehouses that back on the river and there are barges unloading at their wharves or emptied and stranded on mud until the tide rises.

The statistics are impressive. In 1924, there were 103,141 tons of butter, 37,397 tons of bacon, 36,838 tons of eggs, 23,851 tons of cheese and 16,489 tons of lard landed. Sidney Elmer, as a 14-year-old employee of E.M. Denny & Co., later wrote about the horse vans leaving Tooley Street bound for one of London's markets:

> Frequently the street was ankle deep in manure – messy in winter and smelly in summer. A typical load for one of these horse vans was 40 bales of bacon weighing between four and five tons. With its cobble stones and steep incline Duke Street Hill presented a slippery slope for the horses. So the Hay's Wharf people … had a third horse stationed at the foot of the hill with a boy in charge, to harness to the front of any van in difficulty.

Food Processing & Manufacture

Food was both processed and manufactured in Bermondsey. Processing included tea packing, grain and spice milling, margarine production and canning. Biscuits, cakes, ice cream, sauces and pickles were manufactured.

Biscuits

Biscuit manufacture is a distinctly British institution, mainly because the climate is particularly suitable for producing ideal wheat flour. Peek Frean & Co. of Mill Street, Bermondsey, date from when Mr Peek, a tea merchant in the City, invited George Hender Frean, a miller and ships' biscuit maker from the West Country, to manage a biscuit factory he had set up for his two sons. In the event his sons had little or no interest, but Peek Freans nevertheless went from strength to strength. In the early nineteenth century, biscuits were barely edible, but Peek Freans were to change all that. In 1865, they introduced their 'Pearl' biscuit, the pioneer of the modern biscuit – crisp, crumbly and very palatable. They swept the country and, moreover, played a momentous part in feeding the starving of Paris after the Franco-Prussian War. The French government bought 16,000 tons of Peek Freans' best biscuits in 1876 to relieve the siege of Paris. By this time, fire had ravaged the Mill Street premises and the firm expanded to Drummond Road to the south-east. The French order kept Peek Freans going day and night, and the French did not forget: years later, 'Biscuit Town', as it was sometimes known, was visited by the French Ambassador, M. Corbin, and a group of French pensioners, including two who had been besieged sixty-three years before. Henry Mayhew, the well-known social researcher of London life, paid a visit as well and paints a pleasant picture:

> There was a lovely smell of the sweetest flow pervading the immense room which was filled with the haze of the powdery white particles, till the atmosphere seemed to be as gauzy as a summer morning mist. The bakery itself was about as long as between the decks of an Indianman, but as lofty as a railway terminus and heaven knows how many of these same biscuit making machines were at work at the same time.

In 1921, Peek Freans amalgamated with Huntley & Palmers and formed Associated Biscuit Manufacturers. Later, in 1969, they joined with Jacobs to become Associated Biscuits Ltd. In 1982, they became part of the American multinational Nabisco. Manufacture ceased in Bermondsey in 1989.

The biscuit manufacturers McVitie & Price were founded in Scotland, and by 1900 their sales in London had increased tenfold in four years, prompting them to open a factory in Waxlow Road, Harlesden. It was under the direction of Sir Alexander Grant who, amongst other things, helped to build the National Library of Scotland and controversially provided a Rolls-Royce motor car for Labour Prime Minister Ramsay MacDonald. In the late 1940s, the company were making thirty varieties of biscuit and employing 2,000 workers.

McVitie's biscuit factory, Harlesden.

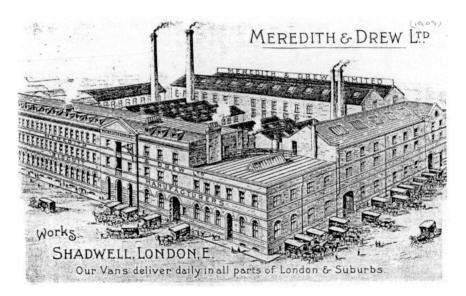

Meredith & Drew biscuit factory, Shadwell.

In 1948, McVitie & Price combined with Macfarlane Lang to form United Biscuits. James Lang had opened a shop and bakehouse in Glasgow and was soon joined by his nephew, John Macfarlane. Once again, because of demand in London, a site in Fulham was acquired where they built the Imperial Biscuit Works. The Gas Light & Coke Co. bought the site in 1929, by which time Macfarlane Lang had moved to Osterley, Middlesex. Goods were transported to and fro by a train which came to be known as the Biscuit Express.

The firm of Meredith & Drew began when William Meredith opened a factory in King Street, Commercial Road, Shadwell. His assistant was William

George Drew. The pair quarrelled and went their separate ways but later made up and merged again to form Meredith & Drew, the order of names decided by the toss of a coin. The firm held the royal warrant to provide biscuits to Queen Victoria, Edward VII and George V, and also the House of Lords. In 1909, an associate of the firm was in Paris and witnessed a street vendor selling 'perles de Paris', a wafer made from potatoes fried in oil. He immediately brought the Frenchman back to London with his recipe and so was born the humble potato crisp. In 1966, Meredith & Drew merged with United Biscuits.

Smith's Potato Crisps was founded by Francis Leigh Smith in his garage at Cricklewood. In 1927, the company moved to Brentford on the ribbon development of the Great West Road. Smith's is now owned by PepsiCo, which now favours the brand made famous by Gary Lineker.

Jam

W.P. Hartley, jam maker of Liverpool, built a large factory at Green Walk in Bermondsey. The premises were opened in 1902 by local MP H.C. Cust, in the presence of the mayor. Hartley welcomed his guests and assured them, to much applause, that all the raspberries were grown in England and that he never used foreign fruit. The *South London Press* of 29 June 1901 carried a story about Mr Hartley. The article – conscious, no doubt, of keeping trade secrets – describes how gooseberries were prepared 'in a machine which is one of the most interesting ever devised' to remove their hairs. Another 'cleverly designed machine' automatically peeled oranges for marmalade preparation. The company also made jelly, lemon curd, coffee essence and mince pies.

The September 1932 issue of Hartley's staff magazine tells how jam was made. Strawberries arrived in late June and the first job was to remove the stalks. Every strawberry was 'plugged' by hand by teams of girls and then inspected before being placed on a travelling conveyor belt to be washed. Perfect preservation depends on correct boiling with the appropriate amount of sugar: too low a temperature and the jam will ferment, while over-boiling leads to a dense and unattractive mass. The still-warm jam was then manually transferred to sterilised glass jars.

C. & E. Morton's jam factory was on the Isle of Dogs in Millwall. The football club that now plays in New Cross was founded at Morton's jam factory in 1885, as Millwall Rovers, by a group of workers of Scottish descent (hence their colours of blue and white). They lost their first game 5-0 to Leytonstone, but went on to achieve a twelve-game unbeaten run. Their first ground was near to where West Ferry Printers were once situated, close to the entrance to the Millwall Dock, and it was only in 1910 that they relocated across the river to The Den.

The firm of Clarke, Nickolls & Coombs were confectioners and jam makers. They came to Hackney Wick in the late nineteenth century and in 1910 were in Wallis Road, near the Lea Navigation. A very go-ahead company, they introduced profit-sharing in 1890 and were probably the biggest employer in the area. They are known for Clarnico Mints (from *Cla*rke, *Ni*ckolls & *Co*ombs). They moved in 1955 to nearby Waterden Road and were later taken over by Trebor Sharps.

James Keiller made marmalade in Silvertown when he arrived in the area from Scotland in 1880. He built a large factory at Tay Wharf. The firm was later taken over by Crosse & Blackwell and then in 1969 by Nestlé.

Flour Mills

McDougalls started in Manchester in 1846, when Alexander McDougall opened a chemical works. He was soon to discover a new baking powder, and his son, Arthur, and two of his brothers set up a fertilizer plant in Millwall in 1869 on the south quay of Millwall Dock. In 1879, McDougalls self-raising flour was born, soon to become a household name. The plant suffered a disastrous fire in 1899 when, despite the attentions of twenty-five fire engines from all over London, the original mill burnt down. It was replaced by the Wheatsheaf Mills a couple of years later. In 1935, enormous concrete silos were erected and became a familiar landmark. The company was later to become part of Rank Hovis McDougall. The Millwall Granary closed in 1982.

In Bromley-by-Bow, on the banks of the Lea, were the Sun Flour Mills. They fell victim to a serious fire in 1952.

Thomas Richard Allinson believed in the benefits of wholemeal bread and founded the Natural Food Co. in Bethnal Green in 1892. He took over the Cyclone Flour & Meal Co. at 21 Patriot Square and began grinding wheat between millstones rather than steel rollers. By this means, essential nutrients and fibre were retained in the bread. For his radical views, Allinson fell foul of the medical establishment. Nevertheless, he was producing 11 million loaves every year by 1911. Offices were built in Cambridge Road, and Allinsons went on to become the largest wholemeal milling company in the country.

There were three large mills beside the Royal Victoria Dock. One, built by the Co-Operative Wholesale Society, occupied 5 acres. Premier Mill, of Joseph Rank Ltd, was nearby. It became Rank Hovis McDougall. There was also Millennium Mill by W. Vernon & Sons. It was destroyed in the Silvertown explosion and rebuilt as Spillers.

Sugar

Sugar is refined from either sugar cane or sugar beet. The end product, sucrose, is the same from both sources. Sugar cane grew in the east and the first person from the western world to discover it was Alexander the Great. Juice, contained in the stalks of the tall reeds, can be extracted and crystallised to a brown, sticky substance. The Romans were aware of it, and Venice became the centre of the sugar trade in the Middle Ages. Before the sixteenth century, honey was the only form of sweetener available to Londoners, who went to the aptly named Honey Lane, off Cheapside, to buy their supplies. Refining was begun in London in 1544 by Cornelius Bussine, even though, as Stow tells us, Antwerp was its main source: 'There were but two sugar houses; and their profit was but little, by reason that there were so many sugar bakers in Antwerp and sugar came thence better and cheaper than it could be afforded in London.'

Large amounts of sugar were imported to Britain in the eighteenth century from the sugar plantations of the West Indies. The stem of the sugar cane – sometimes up to 2in thick – contains a fibre-like substance within which the sugar is stored. After harvesting the cane, containing about 15 per cent sugar, it is squeezed between rollers to expel the juice. Some impurities are removed and then the juice is evaporated to allow sticky brown sugar to crystallise. It is then shipped to England to be refined.

Bussine would have purified his raw sugar in this way: 'Melt ye sugar to ye degree ye may desire and add lime and bullock's blood to clarify.' In this rather unpleasant process, blood and lime caused impurities to flocculate and rise to the surface of the liquid as a scum which could then be scraped off. The liquor was then filtered and boiled to allow the sugar to crystallise. It was sold as loaves – hence sugar bakers. Many sugar bakers came from Germany and set up in the Stepney area. Also in the same area was the firm of Messrs Fairrie Brothers, with premises behind Whitechapel Church.

Henry Tate, son of a clergyman, was born in Lancashire in 1819. He set up on his own as a grocer and, in 1859, went into partnership with John Wright, a sugar refiner of Manesty Lane, Liverpool. But the poet Tony Tate tells the story better:

Now back in the Georgian era
Up North was a parson named Tate.
He had raised up a Christian family
With Henry as child number eight.

Another son, probably Caleb,
Had gone into grocery trade.
Pa said: 'Better take Henry in with you,
– and mind that he's properly paid.'

For some years the family toiled on
And built up a nice chain of shops.
But Henry kept thinking of sugar
And learning of beet and cane crops.

He said: 'Ee, all yon sugar looks scruffy
– those loaves are all more grey than white.'
So he went to an aged Refiner
Well known as old John Wright.

Tate said: 'I've a mind to be changing,
I'll give all my shops to in-laws.
I've a bit of brass saved in my wife's stocking
And I don't mind a business like yours.'

> So Henry and Wright set up in business
> And made quite an impact no doubt.
> Till old man was called to his fathers
> And Henry had business for nowt.

The refinery's name was changed to Henry Tate & Sons in 1869 and three years later, Henry opened the Love Lane Refinery in Liverpool. But the call of London was strong and in the mid-1870s Henry Tate bought land in Silvertown, previously occupied by a shipyard, and built a sugar refinery there. The poet continues:

> Then he called seven sons into office
> And said: 'Four up here will remain.
> But Edwin and two go to London.'
> Ted said: 'Is the man really sane?
> He's bought up a marsh and a gas works
> At least seven miles out of town.
> We'll either go down with swamp fever
> Or whole ruddy workforce will drown.'

To replace the somewhat inconvenient procedure of producing sugar loaves, Henry Tate was quick to exploit a German patent which enabled sugar to be produced in cube form, and it was at Silvertown that Tate made his famous sugar 'cube'. Henry Tate is remembered for his collection of paintings and bequest of money, which made possible the famous Tate Gallery:

> Sir Henry Tate, Bart
> Was a lover of art
> Whom the nation should thank
> For his gallery on Millbank.

Abram Lyle was a Scot who started his working life in a lawyer's office. He later joined his father's cooperage business, then entered shipping and was involved, amongst other things, in transporting sugar. This led to his interest in sugar refining and in particular golden syrup.

> The first Lyle known for our purpose
> Was Abram, the 'Pirate King'
> Who spent all his time up in Greenock
> At what's now called 'doing his thing'.

> His 'thing' was concerned with sea-faring
> And cargoes both arid and wet.
> Till one canny Scot paid in sugar
> And said: 'Juist tak that for ma debt.'

Old Abe called his sons all around him
And asked: 'What's to do wi' yon muck?'
Young Abe said: 'Ye'd better refine it
Or sides of that hold will get stuck.'

So that was the start of his venture
Which prospered and famously grew
On sugar and Lyle's Golden Syrup
(Which Tate's call 'that Devil's Brew').

Henry Tate's sugar refinery, Silvertown, in the 1890s.

Tate & Lyle, Silvertown.

In 1881, Abram Lyle bought land just over a mile upstream of Tate's refinery, near Odams and Plaistow Wharf, and started to make Golden Syrup. The two firms amalgamated in 1921, but strangely enough, the two men never met. The firm's trademark is taken from the Book of Judges, Chapter XIV – the lion that Samson killed full of honey and surrounded by bees, hence the solution to Samson's riddle, 'Out of the strong came forth sweetness.'

Raw sugar arrives at the Silvertown refinery in specially equipped seagoing vessels as moist brown sugar coated with molasses. The first refining process is affination. Warm syrup is added to soften the molasses and the mixture, known as magma, is centrifuged to separate off the syrup. At the carbonation stage the sugar crystals are dissolved in water, lime added and carbon dioxide bubbled through to form chalk, which traps the impurities and is then filtered off. The resulting liquid is passed over charcoal to remove colour and then boiled in vacuum pans to induce crystallisation. After centrifuging, a damp sugar results which is dried in large revolving cylinders in a stream of hot air.

Vinegar

Any American will tell you that fish and chips is the national dish of the British. And fish and chips is not the same without vinegar. The manufacture of wine vinegar began in France in 1670, but as befits a nation of beer drinkers, vinegar manufacture in Britain emerged from the brewing industry. Before the 1860s, vinegar was prepared by allowing alcohol to convert to acetic acid, aided by the action of bacteria, in the open air. It was a lengthy process for the chemical reaction to complete and so a new method was introduced in the late nineteenth century. Due to increased exposure to bacteria, acetification was achieved in five days by circulating the alcohol over birch twigs or beech shavings.

Vinegar was first produced in Castle Street, Southwark, in 1641, and south London became its centre. In the mid-nineteenth century, over half of the country's supply was produced by four companies: Charles W. Potts of Southwark; Sir Robert Burnett of Vauxhall; Henry Beaufoy of South Lambeth; and Slee, Payne & Slee of Horsleydown.

Beaufoys date from the 1740s, when the Quaker Mark Beaufoy opened a vinegar refinery in Cuper's Gardens on the bank of the Thames. It was described by the topographer Thomas Pennant thus: 'The genial banks of the Thames opposite our capital, yield almost every species of wine, and, by a wondrous magic, Messrs Beaufoy pour forth … this ocean of sweets and sours.' When Waterloo Bridge was first proposed (then known as Strand Bridge), Beaufoy was forced to move – but not without compensation of £36,000 – to South Lambeth, between Vauxhall and Stockwell, at a site on Noel Caron's estate.

Slee, Payne & Slee began in 1814, when Noah Slee opened a works in Roper Lane, Southwark. The company merged with Champion & Co. of City Road, Old Street, in 1908 and was later acquired by Crosse & Blackwell, which also absorbed Sarson's. Thomas Sarson had been making vinegar since 1794. In 1893, Sarson's were at 'The Vinegar Works', Catherine Street, City Road, Shoreditch.

Companies merged in the twentieth century and, in 1932, British Vinegars Ltd were established. They were bought by the Swiss Nestlé group in 1979 and traded as Sarson's and Beaufoy's. Manufacture ended in Southwark in 1992.

Crosse & Blackwell

Crosse & Blackwell needed vinegar for their pickles and sauces. The firm can trace their history back to the eighteenth century when Messrs West and Wyatt were trading as London grocers. In 1829, they were bought up by two of their employees, Mr Crosse and Mr Blackwell, and Crosse & Blackwell became producers of canned food. In 1912, Scott of the Antarctic took their canned products on his epic and ill-fated voyage to the South Pole. Years later, in the 1950s, the cans were found with their contents still perfectly wholesome. The original factory was in Charing Cross Road, on the site of the Astoria, but after the First World War they moved to Branston, near Burton upon Trent – hence Branston Pickle. The move was short-lived and soon the firm were back in London, in Crimscott Street, Bermondsey. The company boasted a range larger than their rivals Heinz, and as well as Branston Pickle made piccalilli, Worcestershire sauce, tomato ketchup and salad cream. The Crimscott Street premises closed in the 1960s.

Wall's

Wall's purchased a site at Friars Place, Acton, in 1919, and built a factory to make sausages and pies. They had a slaughter house in Warple Way and pigs were delivered there from the nearby Great Western Railway. Another factory was built in Atlas Road for slaughtering and processing bacon in 1936. After 1956, the Friars Place factory made ice cream and then all meat business was carried out at Atlas Road. At that time, Wall's were employing over 3,000 people.

Milk

George Barham was the first man to import milk into London by train. He was born in 1836, and although he was trained as a carpenter and builder, he followed in his father's footsteps by opening a dairy in 1858 – it was in Dean Street, off Fetter Lane, in the City. Later he opened a new shop at 28 Museum Street, Bloomsbury, conveniently close to King's Cross Station. He called his business the Express Country Milk Supply Co. Its trademark was a locomotive with the nameplate *Express*. At the same time, his father, who had a shop at 272 Strand – supposedly once patronised by Nell Gwynn – changed its name to that of his son's business.

Milk was of doubtful quality at the time. Cows were badly looked after and their milk was frequently distributed in dirty pails. Not only that, it was often diluted with water. Then, in 1865, a cattle virus struck; many cows were destroyed causing a milk famine in London. George Barham took full advantage and began importing milk from Derbyshire via the Great Northern Railway. It was still dark when the milk train drew into King's Cross every morning at 4 a.m. In consultation with the iron maker Thomas Firth, special churns were designed to transport the milk, and help was sought from the brewing trade about how to chill it.

It was around this time that margarine began to appear in the shops. It was developed following the discovery by the French chemist Michel Chevreul of the technique of hydrogenation, whereby solidified animal and vegetable oils could be converted to a butter-like substance. Barham saw this as a threat to his business and pressed the government to pass the Margarine Act of 1887, which decreed that margarine had to be distinguished from butter.

In 1852, the business became the Express Dairy Co. The managing director was Barham's son, Titus, and by that time the company had opened their first tea shop. Meanwhile, Titus' younger brother, Arthur, was running the Dairy Supply Co., a subsidiary of the main company. In 1917, the Dairy Supply Co. combined with Metropolitan & Great Western Dairies of Praed Street, Paddington, and Wiltshire United Dairies of Albert Embankment, Lambeth – both taking their supplies from the Midlands and the South – to form United Dairies Ltd, the forerunner of Unigate. In family businesses members often fall out, and so it was with this one. Arthur expected Titus to retire, enabling him to take over the entire Express business. Titus refused and so the company split, leaving the two brothers in bitter competition with each other.

Milk production was becoming an ever more sophisticated industry. Purity was improved by pasteurisation, whereby bacteria were removed by heating the milk to a temperature some way below its boiling point. Less popular was sterilisation in which milk was treated by boiling, a much harsher treatment, and although it gave the milk a long shelf life, it imparted a different taste. Always innovative, Express were successful in 1965, after nine years of research, in developing long-life milk. It was described as the 'biggest revolution in milk processing since pasteurization'.

Titus Barham had bought 200 acres of land in Finchley in 1880 and immediately stocked it with cattle, calling the place College Farm. He later built a bottling plant, as well as having premises at Claremont Road, Cricklewood, near the railway. There was another plant at Bollo Lane, Acton. In 1969, the company became part of Grand Metropolitan and later, in 1992, became part of Northern Foods. Further reorganisations have taken place since.

Schweppes

Jean Jacob Schweppe was born at Witzenhausen in the Hesse region of Germany in 1740. He started life selling gems and precious trinkets but was always interested in science – it was his scientific curiosity which led him to read Joseph Priestley's *Directions for Impregnating Water with Fixed Air*. Priestley – better known for his discovery and synthesis of oxygen – experimented on many gases including carbon dioxide, known at the time as fixed air. Schweppe, following Priestley's lead, set about inventing an apparatus to aerate water. So was born the soft drinks industry.

Schweppe arrived in England in 1792 and set up shop at 141 Drury Lane, making his aerated water. He later moved to 8 King's Street, Holborn, and then 11 Margaret Street, Cavendish Square. His mineral water soon found favour with the eminent physician Erasmus Darwin, who, on the advice of Matthew Boulton, recommended it for 'stones of the bladder'. Boulton wrote: 'An aerated alkaline water of this kind is sold under the name of factitious Seltzer water by J. Schweppe at 8 King's Street, Holborn, which, I am told, is better prepared than can easily be done in the usual glass vessels.' By 1798, the term soda water was adopted and was taken by 'persons much exhausted by much speaking, heated by dancing or when quitting hot rooms or crowded assemblies'. In those early days the drink was sold in unique egg-shaped bottles that – because they could not stand up – had to be stored on their side. This was deliberate because by so doing the cork was kept saturated with liquid, preventing gas from escaping. Wags of the day called them 'drunken bottles'.

After Jacob Schweppe died, the firm went through a number of owners and in 1831 moved from Margaret Street to 51 Berners Street, where it remained until 1895. Expansion was rapid: factories opened at Hendon and Malvern, also at Hammersmith and Vauxhall, and there was a wharf at Bankside. In 1930, Schweppes acquired a 51 per cent stake in Kia-Ora Ltd, makers of lemon squash. The firm was founded in Australia and took the name Kia-Ora from the Maori for 'good health'. There was a factory in Rushworth Street, Southwark. In 1943, Schweppes took control of all of Kia-Ora.

By this time, the company's headquarters was at Connaught Place, Marble Arch, but in the Second World War a move was made to Tibbett's Corner, Putney Heath, for fear of bombing in central London. Reassurances were also given in the war that they were no longer considered a German company.

The twentieth century saw Schweppes take over or merge with many well-known brands. They came to an agreement with Pepsi-Cola whereby Schweppes would bottle and sell Pepsi in the UK and Pepsi would sell Schweppes in the USA. And through their subsidiary, the Park Bottling Co., they took over the Park Royal factory of Pepsi-Cola.

In 1957, a bid was made for the lime juice manufacturer L. Rose & Co., founded in Leith by Lauchlan Rose as lime and lemon merchants. The firm had profited from the Merchant Shipping Act of 1867, which required all ships to carry lime juice as a preventive against scurvy – hence the term 'limey', the favourite nickname for British sailors in North America and extended in time to include all Englishmen. Rose's had premises in Curtain Road from 1875, but later moved their lime juice – said to be good for curing a hangover – to be made in St Albans. Other acquisitions were to follow, including Chivers of Cambridge; Hartleys of Bermondsey and Liverpool, the jam makers; Typhoo Tea; and Kenco coffee. In 1969, Schweppes merged with the chocolate manufacturer Cadbury, and are now known as Cadbury Schweppes.

Beanz Meanz Heinz

Henry Heinz was born in 1844 in Pennsylvania, the eldest of nine children. His first business enterprise ended in disaster and he was declared an undischarged bankrupt. But that did not put him off for long, for he managed to persuade his brother, John, and cousin, Frederick, to put up the money to start F. & J. Heinz, with himself as the salaried manager. Henry must have done well. In 1876, he bought out John and Frederick and the business became H.J. Heinz & Co. Two years later he was in London. He walked – full of self-confidence, with head held high – into Fortnum & Mason in Piccadilly and offered them 'seven varieties of our finest and newest goods'. He never looked back, for, as he remarked, 'they took the lot'.

To begin with, Heinz operated through British agents – the first was in Haydon Street, Minories, just north of the Tower of London – but then in 1898 he opened his own premises at 99–101 Farringdon Road. Trade increased and Heinz began to search for a suitable factory in London so that he could begin manufacturing in England. A pickle factory was found in Peckham, the premises of Batty & Co. at 127 Brayards Road, and Charles Hellen, his Boston branch manager, was sent across to manage it. Business boomed; cream of tomato soup was introduced to Britain in 1910, and three years later sales had risen to £100,000 per year. By now the Peckham factory was too small for the company's needs and plans were made to move to a 3-acre site in Fulham. But, on consideration, this was thought too small and so Heinz moved to a 20-acre site in Harlesden, bordered by the railway on one side and the canal on the other. Up until 1928, the famous baked beans had been imported from America, but then they began to be produced in Britain and – selling £443,000 worth in year one – were an immediate hit. Cans of the ubiquitous beans were, in fact, donated to Robert Falcon Scott to take with him on his ill-fated expedition to the South Pole. In 1957, they were retrieved from Scott's base camp, opened by his son, Sir Peter Scott, and pronounced, forty-six years later, as 'quite magnificent'.

Sainsbury's

John James Sainsbury's father was a frame and ornament maker living in Oakley Street (now Baylis Road), Lambeth, and it was here that John was born in 1844. In the 1850s, he was living in nearby Short Street and working as a shop assistant in the New Cut, just around the corner. It was when he was working for George Gillett, an oil and colour merchant in Strutton Ground, Victoria, that he met and married Mary Ann Staples. The couple were ambitious and were firm believers in work, order, thrift and self-discipline, and it was these qualities that they brought to their first shop, which they opened in 1869 at 173 Drury Lane, on the corner with Macklin Street.

There were no multiples at the time and John's shop – one of many similar – sold butter, milk, eggs and cheese. In 1876, the enterprising couple opened a branch in Kentish Town at 159 Queen's Crescent, and two years later another branch in the same street. Further stores soon opened at Stepney (67 Watney

Street), Brondesbury, Balham and Lewisham, but the jewel in the crown was the magnificent shop in Croydon. It had much gold lettering, tiles (they were easily washable) and was well lit. Indeed, 'keeping the shop well lit' was a particular requirement of J.J. Sainsbury.

The company's original warehouse was in Kentish Town but by the end of the 1880s, expansion dictated newer and more extensive premises. Cox's Horse Repository, in Stamford Street, supplier of Queen Victoria's first riding horse, came up for sale. There was a clause in the lease restricting use to a horse depository, but Sainsbury overcame this and furthermore obtained the site at a knockdown price, so beginning a long association with Southwark.

Sainsbury recognised the value of the area to the food industry. It was near to the centre of overseas food importation at Hay's Wharf in Tooley Street, close to Smithfield Market, and had good rail links. Opened in 1891, the site proved ideal, so much so that a new factory was opened in 1936, principally for the manufacture of sausages and pies. Designed by Sir Owen Williams (architect of the original Wembley Stadium, the *Daily Express* building in Fleet Street and the Boots factory in Nottingham), it was far ahead of its time. The first building in London to be constructed on the flat slab principle (the weight of the floors is supported by twin rows of internal columns), it was altered to act as office and laboratory accommodation in 1975, and was renamed Rennie House.

In 1950, Sainsbury's experimented with self-service at their Croydon branch in London Road, following a visit to the USA to see such retail outlets in operation. Three years later, the Lewisham self-service store opened, which at that time was the biggest supermarket in Europe.

J. Lyons & Co.

Lyons were once one of the largest food manufacturers in the country. Begun by Joseph Nathaniel Lyons in 1887, they had vast premises in Hammersmith, known as Cadby Hall, having once been occupied by Charles Cadby, a piano manufacturer. Lyons moved there in 1894 and eventually expanded to take up the whole area in Hammersmith Road between Blythe Road and Brook Green. At their peak, 30,000 staff were employed. They also had a chain of tea shops and their well-known multistorey corner houses. Margaret Thatcher was a former employee and applied her chemistry degree to investigate ways of preserving ice cream. Lyons was also famous for innovating the first business computer, LEO (Lyons Electronic Office), developed in 1951 to automate clerical tasks. In the 1970s, Lyons went into decline and were taken over by Allied Breweries.

28 Water Mills & Windmills

Water Mills

Water mills in London seem unlikely to us today, but millers had to grind their corn for London's bakers, and while there were none recorded within the confines of the City walls, there were many elsewhere, particularly along the banks of the River Lea.

There are two mills recorded at Thorney Island on the banks of the Tyburn as it circles around Westminster Abbey. The Abbey Mill was sited just within the gate which led from Dean's Yard to Tufton Street. John Harris knew about it to his cost when, in 1557, he was last seen *extens ebrius* – drunk to us latter-day readers – near to the 'mylne dame'. The unfortunate John fell in and drowned. There was another mill, the Queen's Slaughter House Mill, at the point where Abingdon Street becomes Millbank. Present-day Millbank refers to it.

The rapid flow of the River Fleet was exploited by mill owners. The Knights Hospitaller held land centred on St John's Square and their gardens stretched down to the Fleet (or River Wells as it was sometimes known). Here there were two mills, one belonging to the Knights and the other to the nunnery of St Mary, Clerkenwell. Further downstream, near to Charterhouse Street, an enterprising printer built a mill to spin cotton and wool.

In medieval times, the River Fleet was navigable; stone was unloaded from ships moored on its banks to build Old St Paul's Cathedral. By far the largest mill on its bank was opposite Tudor Street near its exit to the Thames. It was granted by King John to the Knights Templar. However, the mill proved to be a problem for shipping, Henry de Lacy, Earl of Lincoln objecting that ships could not get to the Fleet Bridge (between Fleet Street and Ludgate Hill). He was in little doubt where the blame lay, complaining of a 'diversion of water made by the New Temple for their mill standing without Baynard's Castle'. City dignitaries were called in – the Mayor, Constable of the Tower and Sheriff – to speak with 'honest and discreet men and ensure abuses were righted'.

Just outside the eastern borders of the City there were mills associated with the Hospital of St Katharine, using water from the moat at the Tower of London. One of them got in the way of the plans of William de Longchamps, Chancellor of England in the reign of Richard I, who wished to enlarge the moat. He promptly knocked the mill down, much to the 'discommodity' of the hospital. Edward I, some hundred years later, made up for it by building another mill. There was a further mill near to Traitors' Gate.

There was a group of mills on the banks of the River Lea, described as 'the most remarkable series of mills in Essex'. Flour was provided for the bakers of Stratford, famous in their day for supplying the City of London. To the north, in Leyton, was Temple Mills, first held by the Knights Templar, then the Knights Hospitaller and, after their dissolution, by the Crown. The mills were later used

Three Mills, Bromley-by-Bow.

for a variety of industrial purposes, including boring tree trunks to make pipes for use as water mains, as a brass foundry and for the manufacture of sheet lead.

In the vicinity of Bow Bridge, where Pudding Mill Lane meets Marshgate Lane, there was a mill belonging to the Hospital of St Thomas of Acon. Nearby were City Mill and Fuller's Mill, held by the Cistercian Abbey of Stratford Langthorne. By 1321, it was run by the Bridge House Estates, which received rents that contributed to the maintenance of London Bridge.

Further south is Three Mills which survive – or, strictly speaking, two of the mills do – to this day. The third mill was demolished in the mid-sixteenth century to improve navigation along the Lea. In 1728, the mills were purchased by the Huguenot Peter Lefevre and Daniel Bisson. They were used for a variety of purposes, including a distillery. Following restoration, House Mill and Clock Mill are now owned by the River Lea Tidal Mill Trust and are open to the public at certain times.

There were mills south of the Thames on the rivers Wandle and Ravensbourne. Brook Mills, on the Ravensbourne at Deptford, was once owned by John Evelyn, who lived nearby at Sayes Court. The monks of Bermondsey Abbey owned a mill known as Redriff Mill, which, from its name, was probably in Rotherhithe. It was later called the King's Tide Mill and after the Dissolution was acquired by Henry Reve who built 'le Gone powdermill', the first gunpowder mill in the country to be powered by water. Thomas Brickett later became its tenant and the mill continued in operation until about 1600.

The River Neckinger, a small stream running through Bermondsey, had a mill on its bank, built in 1780, used for paper manufacture and later for the leather industry, Bermondsey's staple industry. There were other mills in the area, near the entrance to St Saviour's Dock and on the site of Courage's Brewery in Shad Thames.

Windmills

Windmills were man's earliest machines and were used to grind corn. The problem was, of course, that when the wind stopped blowing, the miller had to stop grinding his corn. The first windmills were known as post mills and consisted of sails secured on a central post. The whole mill could be rotated to bring the sails into the wind. Great skills in carpentry were required for their construction. Tower mills contained machinery to grind the corn in a stationary tower, as the name implies. They were topped by a cap, to which were fixed the sails, and the cap alone revolved to bring the force of the wind to the sails. Smock mills were variations of tower mills, so called because of their similarity in appearance to a man's smock.

A windmill was first recorded in London in the twelfth century in Clerkenwell. Never as common as water mills, there were nevertheless six at Finsbury Fields, dating from 1549, said to have been built on a mound of high ground formed of rubbish. In the City itself, windmills were sometimes mounted on top of water works and used to pump water, and there was an array of windmills around the Isle of Dogs at Millwall. Eventually, the advent of steam power – see Albion Mills (p. 87) – heralded the demise of windmills.

G.C. Arthur has listed those windmills remaining in Greater London today. There is a tower mill, the Arkley Windmill, at Barnet (Ordnance Survey map ref. TQ 218953), dating from 1830. It continued to work until 1916, when it was bought and restored by Colonel William Booth.

There is a well-known tower mill at Brixton (TQ 305744). It was built in 1816 and ground corn by the power of the wind until 1864. Milling commenced again in 1900 but with steam power, and continued until 1934. The mill was restored in the 1960s by a Lincolnshire firm of millwrights and re-equipped with machinery from a derelict Lincolnshire windmill.

In a back garden at Keston (TQ 415640) is a weatherboarded post mill built in 1716, which last operated in 1878. The Plumstead Windmill (TQ 447779) survives in the garden of a public house. It is a tower mill and dates from about 1830. Standing proudly amongst modern housing is the Shirley Windmill (TQ 355651). It was built in the 1850s, but may contain older parts. It has been well restored and is open to the public at certain times.

The justifiably famous smock mill at Upminster (TQ 557868) is also open to the public. It was constructed in 1803, complete with a bakery next door. Steam power supplemented that of the wind in 1811, and milling continued until a storm destroyed the mill in 1927. In 1960, it stood within a newly created park and was restored by Essex County Council and later by the London Borough of Havering.

There is a post mill at Wimbledon (TQ 230724), also open to public view. It was built by a Roehampton carpenter, Charles March, in 1817, and was later operated by a Mr Dann, a local constable. Wimbledon Common was a favourite venue for duels and Dann was forced to arrest the Earl of Cardigan who was caught duelling with Captain Harvey Tuckett.

★★★

London's Industrial Heritage has been an overview of the many industries which have begun, flourished, declined – and sometimes survived – in London since the Middle Ages. Their influence has shaped the London of today and indeed places further afield. It is a subject with many ramifications and fascinating associations, providing a rich substrate for wider study.

Sources of Reference

The Electricity Industry

Bowers, Brian, *History of Electric Light & Power* (Peter Peregrinus, 1982).

Cochrane, Rob, *Landmark of London: The Story of Battersea Power Station* (CEGB, 1963).

—, *Power to the People: The Story of the National Grid* (CEGB, 1985).

—, *Cradle of Power, The Story of Deptford Power Stations* (CEGB, 1989).

—, *The CEGB Story* (CEGB, 1990).

Hannah, Leslie, *Electricity before Nationalisation* (Macmillan, 1979).

Harris, C., 'Electricity Generation in London', *Geographical Review*, XXXI, 1941, pp. 127–34.

Hennessey, R.A.S., *The Electric Revolution* (Oriel Press, 1972).

Parsons, R.H., *The Early Days of the Power Station Industry* (Cambridge University Press, 1939).

The Times newspaper, 16 December 1878, 21 March 1879, 3 April 1882, 13 April 1882.

Gas

Chandler, Dean and Lacey, A. Douglas, *The Rise of the Gas Industry in Britain* (British Gas Council, 1949).

Everard, Stirling, *The History of the Gas Light & Coke Company 1812–1949* (Ernest Benn, 1949).

Falkus, Malcolm, *Always Under Pressure* (Macmillan, 1988).

Hunt, Charles, *A History of the Introduction of Gas Lighting* (Walter King, 1907).

Mesham, Susan E., *Gas: An Energy Industry* (HMSO, 1976).

South Metropolitan Gas Company: A Century of Gas in South London (1924).

Stewart, E.G., *Town Gas: Its Manufacture and Distribution* (HMSO, 1958).

Williams, Trevor I., *A History of the British Gas Industry* (Oxford University Press, 1981).

Post & Telecommunications

Colby, Reginald, 'The Early Days of London's Telephones', *Country Life*, 17 November 1966.

Durham, John, *Telegraphs in Victorian London* (Golden Head Press, 1959).

Robinson, Howard, *Britain's Post Office: A History of Development from the Beginnings to the Present Day* (Oxford University Press, 1953).

Slater, Ernest, *One Hundred Years: The Story of Henley's* (W.T. Henley's, 1937).

www.siemens.co.uk

Various authors at http://atlantic-cable.com

Water Supply & Sewage Disposal

The bulk of information on London's water supply was taken from the excellent book by H.W. Dickinson, originally written in 1948. Information about sewage is from Stephen Halliday's book. Due acknowledgement is given thus:

Dickinson, H.W., *Water Supply of Greater London* (Newcomen Society, 1954).

Halliday, Stephen, *The Great Stink of London* (Sutton, 1999).

Also:

Essex-Lopresti, Michael, *Exploring the New River* (Brewin Books, 1997).
The Water Supply of London (Metropolitan Water Board, 1961).

The Thames Barrier

Wilson, Ken, *The Story of the Thames Barrier* (Lanthorn, 1984).

Bell Founding

Dodd, George, *Days at the Factories*, Ch. XIV (Charles Knight & Co., 1843).
Tyssen, A.D., 'The History of the Whitechapel Bell Foundry', *Transactions of the London &
    Middlesex Archaeological Society*, 1925, pp. 195–226.
'Whitechapel – Where the Tradition of British Bell Founding Lives On', In-house guide book.

Candle Making

Still the Candle Burns (Privately printed for Price's Patent Candle Co., 1972).
www.prices-candles.co.uk/history/history.asp

The Chemical Industry

Anon., *150 Years of Paint and Varnish Manufacturing* (Thomas Parsons & Sons Ltd, 1952).
—, *New Scientist*, 30 July 1964.
—, *Howards 1797–1947* (Privately printed for Howards & Sons, 1947).
—, *The Story of the British Xylonite Company Ltd 1877–1937* (1937).
Berger, Thomas B., *A Century and a Half of the House of Berger* (Waterlow & Sons, 1910).
Campbell, W.A., *The Chemical Industry* (Longman, 1971).
Carless, Capel & Leonard Records, Hackney Archives, ref. D/B/CCL.
Cripps, Ernest C., *Plough Court: The Story of a Notable Pharmacy, 1715–1927* (Allen &
    Hanburys, 1927).
Garfield, Simon, *Mauve* (Faber & Faber, 2000).
Graham, Patrick, 'Kemball Bishop & Co.', in *London Industrial Archaeology*, No 9, GLIAS,
    2008.
McDonald, Donald, *Percival Norton Johnson* (Johnson Matthey & Co., 1951).
Marshall, Geoff, *London's Docklands: An Illustrated Guide* (The History Press, 2008).
Richmond, Lesley, Stevenson, Julie and Turton, Alison (eds), *The Pharmaceutical Industry:
    A Guide to Historical Records* (Ashgate, 2003).
Sainsbury, Frank, *West Ham 1886–1986* (London Borough of Newham, 1986).
Slinn, Judy, *A History of May & Baker 1834–1984* (1984).
Taylor, Rosemary and Lloyd, Christopher, *The East End at Work* (Sutton, 1999).
Thompson, F.M.L., 'John Bennet Lawes', *Dictionary of National Biography*, vol. 32 (Oxford
    University Press, 2004).
Various authors, *Perkin Centenary London: 100 Years of Synthetic Dyestuffs* (Pergamon, 1958).
www.bjn-paint-reunion.co.uk
www.morgancrucible.com

Clockmaking

Anon., *A Hundred Years of Time: The Story of Baume & Co.* (Watchmakers, 1949).

Loomes, Brian, *The Early Clockmakers of Great Britain* (NAG Press, 1981).

McKay, Chris, *John Moore & Sons of Clerkenwell, Turret Clock Makers* (Pierhead Publications, 2002).

Mercer, Tony, *Mercer Chronometers: History, Maintenance and Repair* (Mayfield Books, 2003).

Page, William (ed.), Victoria County History, *Middlesex*, vol. 2, 1911.

Shears, P.J., in *Proceedings of the Huguenot Society of London*, vol. XX, No 2, 1960, p. 188.

White, T.E., 'A Clerkenwell Tour of Fifty Years Ago', *Antiquarian Horology*, vol. X, 1976/78, p. 340.

Engineering

Anon., *A Brief Account of Bryan Donkin FRS and the Company He Founded 150 Years Ago 1803–1953* (Bryan Donkin Co., 1953).

Barnes, C.H., *Handley Page Aircraft since 1907* (Putnam, 1987).

Barnes, C.H. and James, D.N., *Shorts Aircraft since 1900* (Putnam, 1989).

Bentley, W.O., *An Illustrated History of the Bentley Car 1919–1931* (George Allen & Unwin, 1964).

Buchanan, D.J., *The J.A.P. Story 1895–1951* (J.A. Prestwich & Co., 1951).

Buchanan, R. Angus, 'Henry Maudslay', *Dictionary of National Biography*, vol. 37.

Davies, R.E.G., *A History of the World's Airlines* (Oxford University Press, 1964).

Dickinson, H.W., 'Joseph Bramah and his Inventions', *Transactions of the Newcomen Society*, 1941/2, vol. 22, p. 169.

Jackson, A.J., *De Havilland Aircraft since 1909* (Putnam, 1987).

Marshall, Geoff, *London's Docklands: An Illustrated Guide* (The History Press, 2008).

Martin, J.E., *Greater London: An Industrial Geography* (G. Bell & Sons, 1966).

Nixon, St John C., *The Antique Automobile* (Cassell & Co., 1956).

Petree, J. Foster, 'The Lambeth Works of Maudslay, Sons & Field', *Engineer*, 8 June 1934, p. 585.

—, 'Maudslay, Sons & Field as General Engineers', *Transactions of the Newcomen Society*, vol. 15, 1934, p. 39.

Reilly, Leonard and Marshall, Geoff, *The Story of Bankside* (London Borough of Southwark, 2001).

Rolt, L.T.C., *Tools for the Job: A Short History of Machine Tools* (Batsford, 1965).

Smith, Dennis, *Hydraulic Power* (Greater London Archaeology Society Publication, 1978).

Taylor, H.A., *Fairey Aircraft since 1915* (Putnam, 1974).

Wailes, Rex, 'George Wailes & Co.', *Transactions of the Newcomen Society*, vol. 47, 1975/6.

Whyte, Adam Gowans, *Forty Years of Electrical Progress: The Story of the G.E.C.* (Ernest Benn, 1930).

Wilson, Charles and Reader, William, *Men and Machines: A History of D. Napier & Son* (Weidenfeld & Nicolson, 1958).

www.aecsouthall.co.uk/

http://middx.net/aec/

Footwear

Dobbs, Brian, *The Last Shall Be First: The Colourful Story of John Lobb, the St James's Bootmakers* (Elm Tree Books, 1972).

Hall, P.G., 'The East London Footwear Industry', East London Papers, vol. 5, 1962, p. 3.

Furniture

Kirkham, Pat, Mace, Rodney and Porter, Julia, *Furnishing the World: The East End Furniture Trade 1830–1980* (Journeyman, 1987).

Symonds, R.W., 'City of Westminster and its Furniture Makers', *The Connoisseur*, vol. C, 1937, pp. 3–8.

Glass

Bowles, William Henry, *History of Vauxhall and Ratcliff Glass Houses and their Owners 1670–1800* (Privately printed, 1926).

Buckley, Francis, *Old London Glasshouses* (Stevens, 1915).

'Powells': The Whitefriars Studios', *Journal of the British Society of Master Glass Painters*, XIII, 1960, p. 321.

Leather

Bardens, Dennis, *Everything in Leather: The Story of Barrow, Hepburn & Gale* (Grange Mills, 1948).

Bevington, Geoffrey, *Bevington & Sons, Bermondsey 1795–1950: A Chronicle* (Museum of London, 1993).

Hunting, Penelope, *The Leathersellers' Company: A History* (Leathersellers' Company, 1994).

Marshall, Peter, 'West Bermondsey: The Leather Area; An Industrial Archaeology Walk' (Privately published leaflet, 1992).

Match Making

Beaver, Patrick, *The Match Makers* (Henry Melland, 1985).

Charlton, John, *It Just Went Like Tinder: The Mass Movement and New Unionism in Britain* (Redwords, 1999).

Sainsbury, Frank, *West Ham 1886–1986* (London Borough of Newham, 1986).

Paper, Printing, Newspapers & Bank Notes

Paper

Cossons, Neil, *The BP Book of Industrial Archaeology* (David & Charles, 1975).

Crocker, Alan, 'The Wandsworth Paper Mills', *The Wandsworth Historian*, No 5, 1986.

Jenkins, Geraint, *The Craft Industries* (Longman, 1972).

Overton, J., 'A Note on the Technical Advances in the Manufacture of Paper', in C. Singer, *The History of Technology* (Oxford, 1984), p. 411.

Owen, Roderic, *Lepard & Smiths Ltd 1757–1957: A Desk Book of Paper and Printing* (Privately printed, 1957).

Taylor, Rosemary and Lloyd, Christopher, *The East End at Work* (Sutton, 1999).

Printing

Anon., *Fifty Years of Numbering Machines* (W. Lethaby & Co. Ltd, 1963).

Liveing, Edward, *The House of Harrild 1801–1948* (Harrild & Sons, 1948).

Mole, Tom, 'Stanhope Press', *The Literary Encyclopaedia*, 2002.

Plomer, Henry R., *A Short History of English Printing, 1476–1900* (Kegan Paul, Trench, Trubner & Co. Ltd, 1915).

Reilly, Leonard and Marshall, Geoff, *The Story of Bankside* (London Borough of Southwark, 2001).

Sainsbury, Frank, *West Ham 1886–1986* (London Borough of Newham, 1986).

Twyman, Michael, *The British Library Guide to Printing: History and Techniques* (The British Library, 1998).

Page, William (ed.), Victoria County History, 'Industries', *Middlesex*, vol. 2, 1911.

Newspapers

Barson, Susie and Saint, Andrew, *A Farewell to Fleet Street* (Historic Buildings and Monuments Commission for England, 1988).

Boyce, D. George, 'William Maxwell Aitken', *Dictionary of National Biography*, vol. 1 (Oxford University Press, 2004).

—, 'Alfred Charles Harmsworth', *Dictionary of National Biography*, vol. 25 (Oxford University Press, 2004).

Lake, Brian, *British Newspapers: A History and Guide for Collectors* (Sheppard Press, 1984).

Bank Notes

MacKenzie, A.D., *The Bank of England Note: A History of its Printing* (Cambridge University Press, 1953).

—, *The Later Years of St Luke's Printing Works* (Henry Loveridge Chadder, 1961).

www.bankofengland.co.uk/banknotes

Pottery

Adams, E. and Redstone, David, *Bow Porcelain* (Faber & Faber, 1981).

Bellamy Gardner, H., 'Further History of the Chelsea Porcelain Manufactory', *Transactions of the English Ceramic Circle*, No 9, 1942, p. 136.

Bimson, M., 'John Dwight', *Transactions of the English Ceramic Circle*, No 5, 1961, p. 95.

Cooper, C., 'Old London Potteries', *The Gentleman's Magazine*, August 1892, p. 120.

Eyles, D., *Royal Doulton 1815–1965* (Hutchinson, 1965).

Garner, F.H., 'London Pottery Sites', *Transactions of the English Ceramic Circle*, 1946, p. 179.

Page, William (ed.), Victoria County History, 'Fulham Stoneware', *Middlesex*, vol. 2, 1911.

Shipbuilding

Banbury, Philip, 'Vanished Slipways', *Sea Breezes*, March 1962, p. 190.

Barnaby, K.C., *100 Years of Specialized Shipbuilding & Engineering* (Hutchinson, 1964).

Brown, Angela and Coverson, Ron, 'History of London Yard', www.londonyard.com

Hostettler, Eve, 'Shipbuilding and Related Industries on the Isle of Dogs', in *Dockland: An Illustrated Historical Survey of Life and Work in East London*, S.K. Al Naib (ed.) (NELP & GLC, 1986).

MacDougall, Philip, 'The Royal Dockyards of Woolwich and Deptford', in *Dockland: An Illustrated Historical Survey of Life and Work in East London*, R.J.M. Carr (ed.) (NELP & GLC, 1986).

Mackrow, G.C., *Thames Ironworks Gazette*, No 1, January 1895.

Marshall, Geoff, *London's Docklands: An Illustrated Guide* (The History Press, 2008).

Pollard, S., 'The Decline of Shipbuilding on the Thames', *Economic History Review*, 2nd Series, vol. III, 1950, p. 72.

Simms, Helen, 'Mr Batson's Yard', *The Mariner's Mirror*, vol. 57, 1971, p. 371.

Stansfeld-Hicks, C., 'Shipbuilding on the Thames', *PLA Monthly*, 1927, Two articles, pp. 131, 219.

—, 'Shipbuilding on the Thames', *PLA Monthly*, 1928, Two articles, pp. 79, 361.

Taylor, Rosemary and Lloyd, Christopher, *The East End at Work* (Sutton, 1999).

Lady Yarrow, *Alfred Yarrow: His Life and Work* (Edward Arnold, 1923).

Textiles

Baker, T.F.T. (ed.), Victoria County History, *Middlesex*, vol. 10, Hackney.

Clapham, J.H., 'The Spitalfields Acts 1773–1824', *The Economic Journal*, vol. XXVI, 1916, p. 459.

Flanagan, J.F., *Spitalfields Silks of the 18th and 19th Century* (F. Lewis Ltd, 1954).

Haynes, Alan, 'The Mortlake Tapestry Factory 1619–1703' in *History Today*, vol. 24, 1974, p. 32.

Page, William (ed.), Victoria County History, *Middlesex*, vol. 2, 1911.

Sainsbury, Frank, *West Ham 1886–1986* (London Borough of Newham, 1986).

Schmiechen, James A., *Sweated Industry and Sweated Labour* (Croom Helm, 1984).

Taylor, Rosemary and Lloyd, Christopher, *The East End at Work* (Sutton, 1999).

Thornton, Peter and Rothstein, Natalie, 'The Importance of the Huguenots in the London Silk Industry', *Proceedings of the Huguenot Society*, vol. 20, 1959, p. 66.

Other Establishments & Trades

Anon., 'Bygone Bermondsey', *Bermondsey Bulletin*, September 1954.

Bowbelski, Margaret, 'The Royal Small Arms Factory', *Edmonton Hundred Historical Society Occasional Paper*, New Series, No 35, 1977.

Cottesloe, Lord, 'Notes on the History of the Royal Small Arms Factory, Enfield Lock', *Journal of Society for Army Historical Research*, vol. XII, 1933.

Hayes, David, 'Carreras: family, firm and factory', *Camden History Review*, vol. 27, 2003, p. 30.

Hogg, O.F.G., 'The Development of Engineering at the Royal Arsenal', *Transactions of the Newcomen Society*, vol. 32, 1959/60, p. 29.

Lief, Alfred, *The Firestone Story* (Whittlesey House, 1951).

Marshall, Geoff, *London's Docklands: An Illustrated Guide* (The History Press, 2008).

Masters, Roy, *The Royal Arsenal, Woolwich* (Sutton, 1997).

Negretti & Zambra Centenary, 1850–1950, www.negrettiandzambra.co.uk

Pam, David, *The Royal Small Arms Factory Enfield and Its Workers* (David Pam, 1998).

Reilly, Leonard and Marshall, Geoff, *The Story of Bankside* (London Borough of Southwark, 2001).

The Royal Mint: An Outline History (HMSO, 1967).

Sainsbury, Frank, *West Ham 1886–1986* (London Borough of Newham, 1986).

Taylor, E. Wilfred and Simms Wilson, J., *At the Sign of the Orrery: The Origins of the Firm of Cooke, Troughton & Simms Ltd* (Privately published, 1960).

Taylor, Rosemary and Lloyd, Christopher, *The East End at Work* (Sutton, 1999).

Thomas, David, *Camden History Review*, No 94, March 1986.

Under Eight Monarchs: C.W. Martin & Sons Ltd 1823–1953 (1953).

Watson, J.N.P., 'Long Room – Great Guns: The Story of Purdey's', *Country Life*, 19 May 1977.

Weinstein, Rosemary, *SLAS News*, March 2003.

Whitbourn, Frank, *Mr Lock of St James's Street* (Heinemann, 1971).
www.purdey.com

Canals

Compton, Hugh J., *The Oxford Canal* (David & Charles, 1976).
Denney, Martyn, *London's Waterways* (Batsford, 1977).
Essex-Lopresti, Michael, *Exploring the Regent's Canal* (Brewin Books, 1994).
Faulkner, Alan H., *The Grand Junction Canal* (David & Charles, 1972).
Marshall, Geoff, *London's Docklands: An Illustrated Guide* (The History Press, 2008).
Pratt, Derek, *Discovering London's Canals* (Shire, 1981).

The Docks

Broodbank, Sir J., *History of the Port of London* (Daniel O'Connor, 1921).
Colquhoun, P., 'Treatise on the Commerce and Police of the River Thames', 1880.
Douglas Brown, R., *The Port of London* (Terence Dalton Ltd, 1978).
Foster, J., *Docklands: Cultures in Conflict, Worlds in Collision* (UCL Press, 1999).
Greeves, Ivan S., *London Docks 1800–1980* (Thomas Telford Ltd, 1980).
Hovey, J., *A Tale of Two Ports: London and Southampton* (Industrial Society, 1990).
Marshall, Geoff, *London's Docklands: An Illustrated Guide* (The History Press, 2008).
Milne, Gustav, *The Port of Roman London* (Batsford, 1985).
Pudney, J., *London's Docks* (Thames & Hudson, 1975).
Vaughan, W., 'On Wet Docks, Quays and Warehouses for the Port of London, with Hints Respecting Trade', 1793.

Railways

Allen, Cecil J., *The Great Eastern Railway* (Ian Allan, 1975).
Barker, Theo, *Moving Millions* (London Transport Museum, 1990).
Camp, John, *Discovering London Railway Stations* (Shire, 1969).
Davies, R. and Grant, M.D., *London and Its Railways* (David & Charles, 1983).
Garland, Ken, *Mr Beck's Underground Map* (Capital Transport, 1994).
Howson, H.F., *London Underground* (Ian Allan, 1981).
Jackson, Alan A., *London's Termini* (David & Charles, 1969).
Rose, Douglas, *The London Underground: A Diagrammatic History* (Douglas Rose/Capital Transport, 1980).
Wolmar, Christian, *The Subterranean Railway: How the London Underground was built and how it changed the city forever* (Atlantic Books, 2005).

Brewing & Distilling

Barnard, Alfred, *Noted Breweries of Great Britain and Ireland*, 4 vols (Sir Joseph Causton, 1891).
Cornell, Martyn, 'Three Men Holding Me Down: A History of Truman, Hanbury, Buxton & Co.', *Journal of The Brewery History Society*, No 57, September 1889, pp. 4–12.
Corran, H.S., *A History of Brewing* (David & Charles, 1975).
Gold, Alec, *Four-in-Hand: A History of W. & A. Gilbey Ltd 1857–1957* (W. & A. Gilbey, 1957).
Hardinge, G.N., *The Development and Growth of Courage's Brewery* (Jordan-Gaskell Ltd, 1932).

Janes, Hurford, *Albion Brewery (1808–1958): The Story of Mann, Crossman & Paulin* (Harley Publishing, 1958).

Kinross, Lord, *The Kindred Spirit: A History of Gin and of the House of Booth* (Newman Neame Ltd, 1959).

Mathias, Peter, *The Brewing Industry in England 1700–1830* (Cambridge University Press, 1959).

Putman, Roger, *Beers and Breweries of Britain* (Shire, 2004).

Redman, Nicholas, *Whitbread's at Chiswell Street* (Whitbread, 2000).

Richmond, Lesley and Turton, Alison (eds), *The Brewing Industry: A Guide to Historical Records* (Manchester University Press, 1990).

Ritchie, Berry, *An Uncommon Brewer: The Story of Whitbread 1742–1992* (James & James, 1992).

Serocold, Walter Pearce, *The Story of Watneys* (Watney Combe Reid, 1949).

Strong, L.A.G., *A Brewer's Progress 1757–1957* (Privately printed by Charrington's, 1957).

Wilson, Richard, 'The British Brewing Industry Since 1750', in *The Brewing Industry: A Guide to Historical Records* (Manchester University Press, 1990).

www.gordons-gin.co.uk

www.beefeatergin.com

The Building Industry

Brett-James, N.G., 'A Speculative London Builder of the Seventeenth Century: Dr Nicholas Barbon', *Transactions of the London & Middlesex Archaeological Society*, vol. VI, 1933, p. 110.

Dyos, H.J., 'The Speculative Builders and Developers of Victorian London', *Victorian Studies*, vol. XI supplement, 1968, p. 641.

Page, William (ed.), Victoria County History, *Middlesex*, vol. 2, 1911.

Summerson, John, *The London Building World of the Eighteen-Sixties* (Thames & Hudson, 1973).

—, *Georgian London* (Barrie & Jenkins, 1988).

The Coal Trade

Dale, Hylton B., *Coal and the London Coal Trade* (Howlett & Son, 1912).

Fraser-Stephen, Elspet, *The Story of Charringtons: Two Centuries in the London Coal Trade* (Privately printed, 1952).

Passingham, W.J., *London's Markets* (Sampson Low, Marston & Co., 1935).

Food

Adams, James, *A Fell Fine Baker: The Story of United Biscuits* (Privately printed by Hutchinson Benham, 1974).

Anon., *JS 100: The Story of Sainsbury's* (J. Sainsbury, 1969).

—, *Mr. Cube's Roots: The Story of Sugar* (Tate & Lyle Ltd, 1962).

Baker, T.F.T. (ed.), Victoria County History, *Middlesex*, vol. 10, Hackney, 1995.

Barty-King, Hugh, *Making Provision: A Centenary History of the Provision Trade* (Quiller Press, 1986).

Bird, Peter, *The First Food Empire: A History of J. Lyons & Co.* (Phillimore, 2000).

Cathcart Borer, Mary, *England's Markets: The Story of Britain's Main Channels of Trade* (Abelard-Schuman, 1968).

Cutcliffe, Nick, *100 Years of Progress* (H.J. Heinz, 1986).

Forshaw, Alec and Bergström, Theo, *Markets of London* (Penguin, 1983).

Hall, Sharon, *A Buildings Recording Survey at the Former Sarson's Vinegar Brewery (British Vinegars Ltd), Bermondsey* (Pre-construct Archaeology, 1997).

Hugill, Antony, *Sugar and All That …: A History of Tate & Lyle* (Gentry Books, 1978).

Kerr, Barbara, 'The Beaufoys of Lambeth', *History Today*, vol. 23, 1973, p. 495.

Marshall, Geoff, *London's Docklands: An Illustrated Guide* (The History Press, 2008).

Metcalfe, Colin, *From Crimscott Street: A Personal Memoir of Crosse & Blackwell in Bermondsey in the 1950s* (Privately printed).

Michael Twigg, Brown & Partners, *Hay's Wharf* (Privately printed, 1992).

Morgan, Bryan, *Express Journey 1864–1964: A Centenary History of the Express Dairy Company Limited* (Newman Neame, 1964).

Peek Frean & Co. Ltd: A Hundred Years of Biscuit Making 1857 –1957 (Privately printed, 1957).

Simmons, Douglas, *Schweppes: The First 200 Years* (Springwood Books, 1983).

South London Press, 'The Jam Industry', 29 June 1901.

Southwark Trade Union Support Unit, 'Manufacturing in Southwark', 1986.

Webber, Ronald, 'London's Market Gardens', *History Today*, 1973, p. 871.

Water Mills & Windmills

Arthur, G.C., 'Notes on Windmills in Greater London', in *London Industrial Archaeology*, No 3, 1984.

Reid, Kenneth C., 'The Water Mills of London', *Transactions of the London & Middlesex Archaeological Society*, New Series, vol. XI, part III, 1954, pp. 227–36.

Taylor, Rosemary, *Exploring the East End* (Breedon Books, 2001).

www.housemill.org.uk

Index

If you enjoyed this book, you may also be interested in…

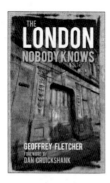

The London Nobody Knows
GEOFFREY FLETCHER

978-0-7524-6199-1

The Little Book of London
DAVID LONG

978-0-7524-4800-5

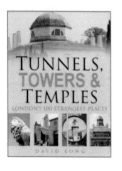

Tunnels, Towers & Temples
DAVID LONG

978-0-7524-4509-7

London's Lost Power Stations & Gasworks
BEN PEDROCHE

978-0-7524-8761-8

Visit our website and discover thousands of other History Press books.

www.thehistorypress.co.uk